Guidebook to
North American
DINOSAURS

According to Created Kinds

© 2014 by Mr. J.D. Mitchell. All rights reserved. Published in the USA by CEC Publications, Gresham, OR.

No part of this publication may be reproduced, stored in a retrieval system, or transmitted in any way by any means – electronic, mechanical, photocopy, recording, or otherwise – without the prior permission of the copyright holder, except as provided by USA copyright law.

ISBN 13: 978-0-615-95291-8
ISBN 10: 0615952917

Contents

Photo and Illustration Credits. vii
Listing of Dinosaur North American Genera ix
Introduction: Dinosaurs—A Different Interpretation. xv

Section I: Dinosaurs and the Bible 1
Section II: Dinosaur Definition. 13
Section III: Dinosaur Footprints. 27
Section IV: Dinosaur Eggs. 37
Section V: Dinosaur Death 45
Section VI: Dinosaur Museums. 51
Section VII: Dinosaur Hunting 65
Section VIII: Dinosaur Paleontologists 71
Section IX: Dinosaur Data by Created Kind 79
    Created Kind #1 – Armor-Backed Dinosaurs 82
    Created Kind #2 – Bipedal Browser Dinosaurs. 88
    Created Kind #3 – Club-Tailed Dinosaurs 95
    Created Kind #4 – Duck-Billed Dinosaurs 98
    Created Kind #5 – Horned Faced Dinosaurs 112
    Created Kind #6 – Horn-Nosed Bipedal Dinosaurs 134
    Created Kind #7 – Killer-Clawed Bipedal Dinosaurs 137
    Created Kind #8 – Lithe, Fast Running Dinosaurs. 142

Created Kind #9 – Long-Necked Big-Clawed Dinosaurs 145
Created Kind #10 – Long-Necked Boxy-Headed Dinosaurs. . . . 149
Created Kind #11 – Long-Necked Slender-Headed Dinosaurs. . 161
Created Kind #12 – Ostrich-Like Dinosaurs. 176
Created Kind #13 – Plate-Backed Dinosaurs 180
Created Kind #14 – Thick-Headed Dinosaurs 189
Created Kind #15 – Tyrant Bipedal Dinosaurs 194

Bibliography 221

Glossary: Dinosaur Terminology. 227

Index 235

Photo and Illustration Credits

ALL OF THE color artistic dinosaur reconstructions and black and white sketches are by creationist artist Marianne Pike.

All other images are by the author **except** for the following:

Section I:
Figure I-1: Walt Brown and Bradley W. Anderson; inset by Donald Wesley Patton, *In the Beginning: Compelling Evidence for Creation and the Flood*, 2008.

Figure I-2: Walt Brown, *In the Beginning: Compelling Evidence for Creation and the Flood*, 2008.

Figure I-3: Alley Oop ©2013 Dist. by Universal Uclick for UFS. Used by permission.

Section III:
Figure III-5: Brent Giles and Paul Abramson, ©2008.

Section V:
Figure V-4: *Science via AP* from www.msnbc.msn.com/id/7285683/

Figure V-5: Dr. Mary Schweitzer in "Dinosaur soft tissue and protein–even more confirmation!" by Carl Wieland, http://creation.com.

Section IX:
Kind #11: *Barosaurus* at American Museum of Natural History, http://newyork.diarystar.com.

Listing of Dinosaur North American Genera

Genus Name	Kind No.	Kind Description	Marker Name
Achelousaurus	#5	Horn-Faced	*Triceratops*
Acrocanthosaurus	#15	Tyrant Bipedal	*Tyrannosaurus*
Alamosaurus	#11	Long-Necked Slender-Head	*Diplodocus*
Albertosaurus	#15	Tyrant Bipedal	*Tyrannosaurus*
Allosaurus	#15	Tyrant Bipedal	*Tyrannosaurus*
Ammosaurus	#9	Long-Necked Big Clawed	*Anchisaurus*
Anatotitan	#4	Duck-Billed	*Edmontosaurus*
Anchiceratops	#5	Horn-Faced	*Triceratops*
Anchisaurus	#9	Long-Necked Big Clawed	*Anchisaurus*
Animantarx	#1	Armor Backed	*Edmontonia*
Ankylosaurus	#3	Club-Tailed	*Ankylosaurus*
Apatosaurus	#11	Long-Necked Slender-Head	*Diplodocus*
Appalachiosaurus	#15	Tyrant Bipedal	*Tyrannosaurus*
Arrhinoceratops	#5	Horn-Faced	*Triceratops*

Continued

Genus Name	Kind No.	Kind Description	Marker Name
Avaceratops	#5	Horn-Faced	*Triceratops*
Barosaurus	#11	Long-Necked Slender-Head	*Diplodocus*
Brachiosaurus	#10	Long-Necked Boxy-Head	*Brachiosaurus*
Brachyceratops	#5	Horn-Faced	*Triceratops*
Brachylophosaurus	#4	Duck-Billed	*Edmontosaurus*
Camarasaurus	#10	Long-Necked Boxy-Head	*Brachiosaurus*
Camposaurus	#8	Lithe, Fast Running	*Coelophysis*
Camptosaurus	#2	Bipedal Browser	*Camptosaurus*
Cedarosaurus	#10	Long-Necked Boxy Head	*Brachiosaurus*
Centrosaurus	#5	Horn-Faced	*Triceratops*
Ceratosaurus	#6	Horn-Nosed Bipedal	*Ceratosaurus*
Chasmosaurus	#5	Horn-Faced	*Triceratops*
Chindesaurus	#8	Lithe, Fast Running	*Coelophysis*
Coelophysis	#8	Lithe, Fast Running	*Coelophysis*
Coelurus	#8	Lithe, Fast Running	*Coelophysis*
Corythosaurus	#4	Duck-Billed	*Edmontosaurus*
Daspletosaurus	#15	Tyrant Bipedal	*Tyrannosaurus*
Deinonychus	#7	Killer-Clawed Bipedal	*Deinonychus*
Diceratops	#5	Horn-Faced	*Triceratops*
Dilophosaurus	#8	Lithe, Fast Running	*Coelophysis*
Diplodocus	#11	Long-Necked Slender-Head	*Diplodocus*
Drinker	#2	Bipedal Browser	*Camptosaurus*

Continued

Listing of Dinosaur North American Genera

Genus Name	Kind No.	Kind Description	Marker Name
Dromaeosaurus	#7	Killer-Clawed Bipedal	*Deinonychus*
Dromiceiomimus	#12	Ostrich-Like	*Ornithomimus*
Edmarka	#15	Tyrant Bipedal	*Tyrannosaurus*
Edmontonia	#1	Armor-Backed	*Edmontonia*
Edmontosaurus	#4	Duck-Billed	*Edmontosaurus*
Einiosaurus	#5	Horn-Faced	*Triceratops*
Eobrontosaurus	#11	Long-Necked Slender-Head	*Diplodocus*
Eucoelophysis	#8	Lithe, Fast Running	*Coelophysis*
Euoplocephalus	#3	Club-Tailed	*Ankylosaurus*
Gastonia	#1	Armor-Backed	*Edmontonia*
Gojirasaurus	#8	Lithe, Fast Running	*Coelophysis*
Gorgosaurus	#15	Tyrant Bipedal	*Tyrannosaurus*
Gryposaurus	#4	Duck-Billed	*Edmontosaurus*
Hadrosaurus	#4	Duck-Billed	*Edmontosaurus*
Hesperosaurus	#13	Plate-Backed	*Stegosaurus*
Hypacrosaurus	#4	Duck-Billed	*Edmontosaurus*
Kritosaurus	#4	Duck-Billed	*Edmontosaurus*
Lambeosaurus	#4	Duck-Billed	*Edmontosaurus*
Leptoceratops	#5	Horn-Faced	*Triceratops*
Maiasaura	#4	Duck-Billed	*Edmontosaurus*
Marshosaurus	#15	Tyrant Bipedal	*Tyrannosaurus*
Monoclonius	#5	Horn-Faced	*Triceratops*
Montanoceratops	#5	Horn-Faced	*Triceratops*
Nanotyrannus	#15	Tyrant Bipedal	*Tyrannosaurus*
Niobrarasaurus	#1	Armor-Backed	*Edmontonia*

Continued

Genus Name	Kind No.	Kind Description	Marker Name
Ornitholestes	#8	Lithe, Fast Running	*Coelophysis*
Ornithomimus	#12	Ostrich-Like	*Ornithomimus*
Orodromeus	#2	Bipedal Browser	*Camptosaurus*
Othnielia	#2	Bipedal Browser	*Camptosaurus*
Pachycephalosaurus	#14	Thick-Headed	*Pachycephalosaurus*
Pachyrhinosaurus	#5	Horn-Faced	*Triceratops*
Paluxysaurus	#10	Long-Necked Boxy-head	*Brachiosaurus*
Panoplosaurus	#1	Armor-Backed	*Edmontonia*
Parasaurolophus	#4	Duck-Billed	*Edmontosaurus*
Parkosaurus	#2	Bipedal Browser	*Camptosaurus*
Pawpawsaurus	#1	Armor-Backed	*Edmontonia*
Pentaceratops	#5	Horn-Faced	*Triceratops*
Plateosaurus	#9	Long-Necked Big-Clawed	*Anchisaurus*
Podokesaurus	#8	Lithe, Fast Running	*Coelophysis*
Prosaurolophus	#4	Duck-Billed	*Edmontosaurus*
Saurolophus	#4	Duck-Billed	*Edmontosaurus*
Sauronitholestes	#7	Killer-Clawed Bipedal	*Deinonychus*
Sauropelta	#1	Armor-Backed	*Edmontonia*
Saurophaganax	#15	Tyrant Bipedal	*Tyrannosaurus*
Sauroposeidon	#10	Long-Necked Boxy-Head	*Brachiosaurus*
Segisaurus	#8	Lithe, Fast Running	*Coelophysis*
Seismosaurus	#11	Long-Necked Slender-Head	*Diplodocus*
Silvisaurus	#1	Armor-Backed	*Edmontonia*

Continued

Listing of Dinosaur North American Genera

Genus Name	Kind No.	Kind Description	Marker Name
Sphaerotholus	#14	Thick-Headed	*Pachycephalosaurus*
Stegoceras	#14	Thick-Headed	*Pachycephalosaurus*
Stegosaurus	#13	Plate-Backed	*Stegosaurus*
Stokesosaurus	#15	Tyrant Bipedal	*Tyrannosaurus*
Struthiomimus	#12	Ostrich-Like	*Ornithomimus*
Stygimoloch	#14	Thick-Headed	*Pachycephalosaurus*
Styracosaurus	#5	Horn-Faced	*Triceratops*
Supersaurus	#11	Long-Necked Slender-Head	*Diplodocus*
Tenontosaurus	#2	Bipedal Browser	*Camptosaurus*
Thescelosaurus	#2	Bipedal Browser	*Camptosaurus*
Torosaurus	#5	Horn-Faced	*Triceratops*
Torvosaurus	#15	Tyrant Bipedal	*Tyrannosaurus*
Triceratops	#5	Horn-Faced	*Triceratops*
Troodon	#7	Killer-Clawed Bipedal	*Deinonychus*
Tyrannosaurus	#15	Tyrant Bipedal	*Tyrannosaurus*
Utahraptor	#7	Killer-Clawed Bipedal	*Deinonychus*
Zuniceratops	#5	Horn-Faced	*Triceratops*

Introduction: Dinosaurs—A Different Interpretation

Why so Many Guidebooks?

THERE HAVE BEEN literally thousands of general-interest books written on the subject of dinosaurs. Among them are probably scores that could be categorized as "guidebooks" where details concerning each type of a number of dinosaurs are laid out in some sort of comparative format. A study of the guidebooks reveals that the dinosaur types are almost always separated according to genus name instead of the more specific species or the more general family classifications. Common examples of these genus names are *Tyrannosaurus*, *Triceratops*, *Edmontosaurus* and *Brachiosaurus*.

An obvious question is why are there so many of these guidebooks? I think the answer is that the general public is interested in dinosaurs and money can be made from books written to satisfy that interest. Whenever a proposed new dinosaur guidebook is envisioned by an author or publisher there must early on in the process be the thought that the new guidebook will offer something that previous guidebooks did not. That something could be new information on a myriad of aspects of the science of paleontology within which the study of dinosaurs resides. Perhaps the driving force is better skeletal reconstructions, a new guidebook format, superior artistic renditions, or more vivid photographs. No matter the reasons, up to this point in time, these guidebooks have all been secularly produced and have in common interpretations based on evolution and millions of years. They have all been based on the same atheistic worldview.

Why this Guidebook?

This guidebook is the first to provide substantial information on North American dinosaurs based on a biblical creationist worldview. Over one hundred genus names have been filtered down to just fifteen created kinds explained in Section IX of this guidebook. The intent has been to provide mostly general up-to-date information on each created kind, but the reader may find that there are facts and photos in this guidebook found in no other.

It is not news to the secular or the creationist paleontologists that the tendency for splitting in the naming of fossil life forms from the rock record has resulted in many bogus genus names for dinosaurs over the years. An in-depth study of the actual reported fossil material reveals that in many cases genus names have been assigned to a very sparse amount of bone fragments or teeth. The evolutionary paradigm, with the underlying assumption of macroevolution, tends to encourage this splitting tendency. Most secular dinosaur guidebooks typically report all the multitude of genus names even though they sometimes also document the lack of supporting fossil material and/or the likelihood of questionable taxonomy for some genus and species names.

For the creationist author, questionable dinosaur taxonomy exacerbates the task of determining which genus names to use in the determination of created kinds. The criteria used in this guidebook are quality and quantity of reported fossil material available. That is, if there are only a few fragments of a particular genus available, then that genus may be ignored. Yes, this is a subjective method, but I have found that in many cases genus names have been assigned based on so little fossil material that the decisions to delete certain ones were not at all difficult. The fact is that the secularly assigned genus names were also made on very subjective bases too.

Variation

As with extant life forms scientists have found that there is much variation within each species of dinosaur. However, in the assignment of names to extant life forms the taxonomist does not have to deal with the following obstacles that dinosaur paleontologists must contend with:

1. They must work with fossil specimens where a large proportion is of poor quality.

Introduction: Dinosaurs—A Different Interpretation xvii

2. The fossils they work with provide a nearly total lack of soft tissue material.
3. In most cases there is available only a very small sample of fossil material.

These obstacles (poor quality, little soft tissue, small samples) along with the pressure to publish materials that are considered valuable to the scholarly community and the influence of the evolutionary paradigm often results in secular paleontology misinterpreting the variation seen in the fossils they study. Whenever variation is discovered it is generally interpreted as being part of macro-evolutionary movement rather than any sort of genetic variation within a kind. For 150 years evolutionists have been unsuccessfully attempting to demonstrate the fact of evolution using the fossils in the rock record. For most of those years attempts were made to use Linnaean taxonomy to show a correspondence of variation in the fossils to Darwinian evolutionary theory. In the study of dinosaurs this attempt has been such a resounding failure that most paleontologists have turned to cladistics analysis to communicate their speculative hypotheses. I think this is a reinforcement of the fact of created kinds and the truth of the separation of life-forms that refutes common descent philosophies.

An Assumption

The creationist paleontologist normally has access to the available fossil material only through secular publications, Internet website information and museum displays. Creationists are not welcome in the "sacred" halls or fossil storerooms of secular universities and natural history museums where the vast majority of recovered dinosaur fossils are to be found. Therefore, in the determination of dinosaur created kinds and the development of this guidebook, I have made the following assumption:

> *If an important dinosaur fossil specimen is of good quality, has been prepared, researched, described, and can be interpreted to fit the philosophy of evolution and millions of years – it, or a cast replica of it, will be put on display somewhere in a major secular museum.*

Upon this assumption I have based the development of much of the material in this guidebook including the determination of which dinosaur genera are valid. In the beginning of the research for this guidebook, the supporting evidence for this assumption had not been totally substantiated. However, my confidence in the truth of the assumption increased with each day of research. Perhaps someday I will be convinced of error in this assumption, but at the time of publication of this guidebook I had no reason to believe the assumption is in any way incorrect.

Museum Displays

My decade-long museum research for the guidebook has resulted in three significant conclusions:

1. Not all dinosaur kinds are equally represented in museum displays. There are a few "popular" kinds that most natural history museums with significant dinosaur displays will always have available for viewing by the public. In fact, there are just three kinds of dinosaurs whose fossil material make up over half of the public museum displays of North American museums. These are the Duck-Billed, Horn-Faced, and Tyrant Bipedal dinosaur created kinds.
2. Fossil displays are often duplicated from museum to museum. This is seen especially with superior quality skulls and skeletal reconstructions. If a reconstructive effort is considered successful it will usually be cast and sold for display at other museums. This practice is also reinforced by independent fossil casting companies in the business to sell museum-quality reconstructions to as many locations as possible.
3. Skeletal reconstructions in museums are often made up of a high percentage of non-bone material. The scarcity of complete articulated dinosaur skeletons results in the need to substitute plaster casts of bones to arrive at a complete reconstructed animal. In many cases these cast parts are from other specimens found at other locations. Most museums will hide this practice by painting the complete reconstruction a common color (often brown, gray or black) so that it is impossible for any other than an expert to discern what is real bone, and what is not.

Introduction: Dinosaurs—A Different Interpretation

As time marches on, those early museum skeletal reconstructions that are made up of mostly bones are oftentimes dismantled. The real bones are all replaced with cast parts molded directly from the bones. That allows paleontologists to save the real bones from further disintegration, modify the skeletal poses to match current ideas, and have better access to the real bones for additional scientific study.

For these reasons the museum-dinosaur-display viewer would be wise to consider the accuracy of each skeletal reconstruction with a healthy amount of skepticism. This skepticism should be even stronger with regard to fully "fleshed-out" animal reconstructions that attempt to show how the dinosaurs looked including their skin. No one knows the color(s) of the skin of dinosaurs and the increasingly common depictions of dinosaurs with feathers are wrought with a very high degree of philosophy-based fantasy.

Fig. 1: Museum Reconstructed Skeletal Dinosaurs on Display

Presuppositions

The reason that secular and creationist interpretations of dinosaur fossils are so different is that each has different presuppositions. I have been observing these two sets of presuppositions in operation for decades as they are applied to the study of paleontology and have summarized them below. Here are the major presuppositions under which secular paleontology operates:

- There is no God, or God is irrelevant.
- Everything came from nothing.
- All life-forms evolved from common ancestors over billions of years.
- Homology (the fact of similarities in life-forms) is sufficient proof for the evolution of all extinct and living life-forms.

Biblical paleontology operates under an entirely different set of presuppositions.

- In the beginning God created everything.
- The Bible is God's true Word to mankind.
- God created in six ordinary days only thousands of years ago.
- A cataclysmic worldwide flood destroyed all land animals and humans, except for those on Noah's Ark some 4,500 years ago.

The biblical set of presuppositions above was foundational in the development of the information in this guidebook, of course. Other important considerations were the biblical statement that God created according to kinds, and a corollary of the fourth biblical presupposition that dinosaur fossils are found in the sedimentary layers of the earth as a result of the worldwide flood that God used to judge mankind.

Assignment of Created Kinds

The assignment of family categories and genus/species names to dinosaurs by secular taxonomists is a highly subjective exercise. For this reason changes and deletions are common as new fossil materials are discovered, errors are found, and different interpretations gain acceptance. The similarities that are found in dinosaurs by secularists are explained to be as a result of common ancestry (sometimes called "descent with modification").

In the development of dinosaur created kinds in this guidebook subjectivity is also an admitted factor. Linnaean taxonomy provided the basis that was used for the taxonomic assignment of kinds. Inherent in this method was an attempt not only to identify similarities, but also an attempt to identify the gaps between kinds. Similarities between animal kinds is understood and explained to be as a result of the work of a Common Designer at the beginning.

From the Bible it can be determined that there were about 1,600 years from the time of creation to the time of the Flood. Initially everything created was "very good," but not long after the creation Adam and Eve sinned against God (the Fall), resulting in death and the curse. From the time of the Fall until the Flood mankind became more and more evil until God stood for it no more. Except for Noah, no one was found righteous, and mankind along with other life-forms was allowed to continue only because of this righteousness and the grace of God. This most ancient historical information is found in the book of Genesis in the Bible.

It is a well-known fact that life-forms can develop genetic variability within a kind very rapidly. A good example of this is the variability developed within the dog kind that has been witnessed and historically documented by mankind. I have concluded that 1,600 years would be sufficient time for genetic variability to modify the created dinosaur kinds to be as found in fossils in the sediments. All of the dinosaur kinds were created in the beginning, lived at the same time, and through genetics over 1,600 years changed so that we find today the variety that was buried in the Flood of Noah's day. The dinosaur fossils we dig up today are in the ground because they were buried during the worldwide flood.

In summary, the assumed factors that drove the development of the kinds of dinosaurs depicted in this guidebook are:

- God created life according to kinds.
- Linnaean taxonomic principles are the best available and most helpful for application to dinosaurs.
- The dinosaur fossils are in the sediments as a result of the worldwide flood.
- Much genetic variation within dinosaur kinds could occur quickly, certainly within 1,600 years.

Section I

Dinosaurs and the Bible

Biblical Timescale

THE HYPOTHESIS PRESENTED in the Introduction is that dinosaurs can be logically interpreted by relying upon the Genesis chapters 1–11 historical narrative in the Bible. The biblical timescale can also be reasonably explained as long as biblical presuppositions are the foundation for the explanation.

The Bible definitely provides an accurate history of the world starting at the very beginning of God's creation. There is no need or justification for inserting extra thousands or millions of years into its historical narrative. Most educated people can understand that it would be possible to calculate the number of years from the present back to the date that their great-grandfather was born knowing certain historical information. Using the example of a man wanting to calculate the time from the present back to the date of birth of his great-grandfather, the calculation would only require the following information:

1. How old is the man today?
2. How old was the man's father when the man was born?
3. How old was the man's grandfather when his father was born?
4. How old was the man's great-grandfather when his grandfather was born?

If the man were 60 years of age at the present and each of his ancestral fathers had their sons 20 years apart the answer to the calculations would be equal to 60 + 20 + 20 + 20 = 120 years.

This is the exact type of chronological information (genealogy) that the Bible records (in Genesis chapters 5 and 11 and following) from the time of creation for over 3,500 years. Other more recent sources can be consulted to develop the time from the end of the biblical chronology to the present. Adding the biblical chronology to the later figures provides the time from creation to the present. The answer to this calculation is always in the ballpark of 6,000 to 7,000 years.

Figure I-1 illustrates how the Bible lays out the time from creation to the death of Joseph in the manner just discussed.

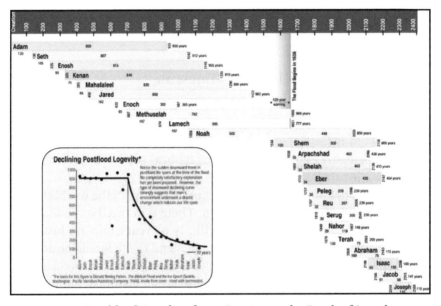

Fig. I–1: Biblical Timeline from Creation to the Death of Joseph

From this genealogy chart we can see that the worldwide flood at the time of Noah took place about 1,656 years after the creation. The worldwide flood was a result of God's judgment on the evil of the world at that time when the only righteous man left on earth was Noah. This cataclysmic Flood caused most of the geology we find on the surface of the earth today and is the main reason we find fossils in the rock record—including dinosaur fossils.

Also in Genesis 1:24–25 we are told that land animals were created on day six of the creation and since dinosaurs are definitely animals, they too were created on day, six. God made man on day six along with the animals (Genesis 1:26–27), and that leaves the unavoidable conclusion that man and dinosaurs lived at the same time for the period from day six up to the Flood—and for some period of time after.

Noah's Ark, Dinosaurs, and Creation Science

Representative dinosaurs were placed on Noah's Ark along with the other land animals and were introduced to the new earth environment after the Flood (Genesis chapter 7). Most, if not all, of these dinosaurs have gone extinct in the approximately 4,500 years since then.

Creationists have logical biblical answers for most of the questions people have for the Flood, the Ark, and the dinosaurs. There are very good creationist resources in the Bibliography that address these questions in great detail, but here for quick reference are just a few of the more common questions and appropriate biblically-based answers:

Question 1: How would there be room for all the dinosaurs on the Ark?

Answer: As will be discovered in Section IX and elsewhere in this book, there were not all that many different kinds of dinosaurs on the Ark. Not all of these dinosaurs were large either, as the average mature dinosaur was probably about the size of a sheep. God brought the animals to Noah and so could have selected younger ones that would have been easier to handle, would have taken up less space, used less food, and have been better suited to repopulate the earth after the one-year-long Flood. It makes good sense that the largest representatives of all animal kinds would have been left off of the Ark. In addition, creation scientists have calculated that there would have been plenty of room for the land animals and birds on the huge 500-foot-long Ark. For great detail on this subject see Woodmorappe's feasibility study that is listed in the creationist bibliography. Figure I–2 puts some perspective on the size of the Ark especially in light of many of the available children's toys sold today that have the giraffe as tall as the Ark.

Question 2: Why didn't the carnivorous dinosaurs and other carnivorous animals eat each other on the Ark?

Answer: The Bible tells us there was food provided for the animals on the Ark. In addition, God could have chosen less aggressive animals from the kinds that had become meat eaters since the Fall and/or God could have either restrained them miraculously (perhaps by inducing a type of hibernation) or commanded them not to eat each other.

Fig. I–2: Drawing of Noah's Ark Sitting in a Football Stadium

Question 3: Why don't we see dinosaurs alive today?

Answer: The earth was much different after the Flood than before the Flood. The dinosaurs were not able to adapt to the more severe environment after the Flood where there followed an Ice Age, many volcanoes, tectonic movements of the crust, dramatic weather extremes, deserts, swamps, canyons and high mountains that probably did not exist before. As humans began to spread out over the earth after the Flood as directed by God, they would see the dinosaurs as food or as a threat to their safety and would kill them as well as destroy the best dinosaur habitats. Therefore, the dinosaurs became extinct kind by kind over time. There is plenty of evidence from written historical records that a significant number of dinosaurs were seen by humans up through at least 1500 AD. Numerous animal pictographs and petroglyphs can be interpreted to depict dinosaurs. In the Bible and in other early literature the dinosaurs were called dragons (see also question 4). It is quite possible some dinosaurs are still alive today in remote areas of the earth such as in deepest Africa.

Dinosaurs and the Bible

Question 4: Why don't we find the word "dinosaur" in the Bible?

Answer: The word "dinosaur" was invented (by Sir Richard Owen of England) in 1841 long after the first English translations of the Bible were introduced. Dinosaurs were called dragons in the early English translations of the Bible. For examples in the King James Bible see Nehemiah 2:13; Psalm 91:13; Isaiah 27:1 and 43:20; Jeremiah 9:11 and 51:34; Ezekiel 29:3; Micah 1:8 and Malachi 1:3. In the book of Job, in chapter 40, is mentioned Behemoth, an animal whose description closely matches the Long-Necked kinds of dinosaurs. The common secular explanation that the Behemoth of the Bible was an elephant or hippopotamus does not meet the easiest of logical tests. Unfortunately, most English Bible translations that were published after the mid-1850s have been influenced by atheistic science and the secular culture to translate the Hebrew word for dragon (*Tannim*) as other animals such as jackals.

Question 5: Hasn't science proved that dinosaurs lived and went extinct millions of years ago?

Answer: The primary way we know that dinosaurs existed is through the fossils that have been found in the sedimentary layers of the earth's crust. There are no dates of any kind on these fossils; no tags or markers. No one is alive that was there when these life-forms were rapidly buried in a watery catastrophe in order to be preserved as fossils. Those who wanted to reject the Bible and minimize the influence of the Christian Church on the world had to make up a story for how these fossils came to be that rejected the biblical account, and so invented the evolution and millions-of-years scenario. With biblical presuppositions it is very reasonable to explain the fossils as a result of the worldwide Genesis flood at the time of Noah. Remember, God was there when the fossils were buried, but secularists were not! Many people have developed opinions about Noah's flood without studying the considerable detail provided in Genesis chapters 7–9 of the Bible.

Question 6: Hasn't science proved that dinosaurs evolved from other animals and then evolved into birds?

Answer: As mentioned already in the Introduction, the variation within kinds of extinct and living life-forms can be extensive, as allowed for by God when he designed life for planet earth at the time of creation. However, true science has not found any example of one kind of life evolving into another kind, either fossil or living. These so-called "transitional" forms are purely in the imaginations of evolutionists. In fact, there is no valid science that would indicate this sort of macroevolution is even possible. All life-form adaptive changes over time seem to be limited by the information (in the DNA) placed in the kind at the time of creation. God created according to kinds and the kinds have propagated (with great although limited variety) within those kinds ever since. It is very clear that information to allow one kind of life to change into another kind cannot come into being from the random interaction of matter or from genetic mutations.

The Bible explains that God created birds on day five (Genesis 1:21) ahead of the animals that were created on day six, so birds could not have evolved from dinosaurs – birds already existed when the dinosaurs were created!

Question 7: Isn't creationism just religion and not science?

Answer: The word "science" means knowledge, and the scientific method uses observation and repeatable experimentation to develop testable hypotheses and theories in order to come to tentative conclusions about the universe. When someone says that science must always reject the origin explanations in the Bible, they are in effect defining science as being confined only to the man-developed, God-rejecting, philosophical view called naturalism. In naturalism God is not allowed to be entered into any consideration for reality no matter what the evidence indicates. No one was there at the beginning (except for God) and so neither evolutionism nor creationism is strictly scientific. That is, there is no scientific experiment any human can devise to test either hypothesis. Both perspectives look at exactly the same evidences but with different and conflicting presuppositions, and that always results in different interpretations for the evidences. The creation science information contained in this dinosaur guidebook is just as scientific, if not more, as the information found in any secular dinosaur guidebook. The advantage of creation science is that it is founded on the reliable Word of the Creator of everything and not on the often-incorrect speculations of fallible men.

Question 8: I'm still confused about just what is science and what is not science. Can you provide me with some examples to help me understand this?

Answer: Yes. Below are listed some activities that paleontologists regularly take part in as a part of their careers. Each activity specifies whether the

Dinosaurs and the Bible 7

activity is true science or just philosophy. This listing is equally applicable to secular or biblical paleontologists:

- Prospect for fossils in the field: Science
- Recover fossils from the earth's crust: Science
- Prepare fossils for study and/or museum display: Science
- Study and describe recovered fossils: Science
- Publish descriptions of methods for recovering fossils: Science
- Publish the geographic locations of fossils: Science
- Publish the physical relationships of fossils to one another as discovered in the sediments (for example, prepare bone charts): Science
- Publish the descriptions of fossils: Science
- Publish information on the variability found in recovered fossils: Science
- Develop and publish "Tree-of-Life" charts: Philosophy
- Develop and publish cladograms for dinosaurs and other life: Philosophy
- Publish speculations about the origin of life-forms found as fossils: Philosophy
- Publish speculations about deep time: Philosophy

Question 9: You haven't explained radiometric dating methods. Hasn't it been scientifically proven the earth is 4.6 billion years old and dinosaur fossils are over 60 million years old?

Answer: A complete explanation of all of the radiometric dating methods and other uniformitarian dating methods is beyond the scope of this dinosaur guidebook. However, here are just a few important facts regarding these methods:

- The radiometric dating methods are based upon unproved and scientifically unprovable assumptions about the past.
- Radiometric dating methods for millions or billions of years cannot be directly used for fossils because these methods are not applicable to the sedimentary rock layers wherein fossils are found. These methods are applicable to igneous, not sedimentary rocks.
- Secular paleontologists seldom use radiometric methods to establish the age of fossils. Usually they use index fossils and

fossil correlation instead, since they have found that radiometric determined dates usually do not match up to the expected already "established" dates for the fossils. Of course, creationists have no reason to trust these methods as useful for dating rocks.
- The results for various radiometric methods do not match each other and never match up to the known ages of igneous rocks. For example, the igneous rocks that resulted from the 1980s eruptions of Mount St. Helens in Washington State were radiometrically dated to be many orders of magnitude older than their actual age.
- The highly speculative ideas of uniformitarianism, evolution, and millions of years were well established long before any of the radiometric methods were developed. As the radiometric methods were developed they were forced to fit within the pre-existent secular long-ages paradigm.
- Many uniformitarian dating methods, including radiocarbon dating, indicate the earth is much younger than millions of years. A number of these methods indicate the earth is less than 10,000 years of age. All of the methods that indicate a young earth are philosophically, not scientifically, rejected by secularists and others who accept naturalism and an old-earth view.
- Fairly recently, dinosaurs and other vertebrate fossils have been recovered with soft tissue inside the bones. Even original dinosaur proteins called collagen and elastin have been discovered in the bones. There is some indication that dinosaur DNA may still be in some bones as well. Secular science has no logical explanation for how any of these materials could last even one million years, let alone the 65 million years generally accepted by evolutionists as when the last dinosaurs went extinct. See Section V for more on this topic.

The question and answer format will conclude at the end of the next topical discussion.

What's Wrong with this Picture?

The Alley Oop comic in Figure I–3 serves as a helpful illustration for ferreting out some misconceptions about the Bible, dinosaurs and ancient man.

Dinosaurs and the Bible 9

Fig. I–3: Alley Oop—What's wrong with this Picture? (Alley Oop ©2013 Dist. by Universal Uclick for UFS. Used by permission. All rights reserved.)

I remember that the Alley Oop comic was a staple of the newspaper to which my parents subscribed when I was a child. For several years it was my favorite comic strip, probably because of the dinosaurs that Alley rode and lived with in the Stone Age kingdom of Moo. Then, as I began accumulating my collection of dinosaur books, people began informing me that the cartoonist V.T. Hamlin had things all wrong. Of course, everyone knew then (as they now should know) that Alley Oop was and is a fictional comic strip and never was intended to be a depiction of reality. Nevertheless, to my "advisors" of the time it seemed of great importance that I understood that dinosaurs died out millions of years before man came onto the scene. So, I was soon convinced to accept the millions-of-years scenario and did so for another thirty years of my life—all because of those secular worldview influences in vogue at the time.

This dinosaur guidebook is written with the biblical view that dinosaurs and mankind lived on earth contemporaneously for thousands of years after the creation. Therefore, V.T. Hamlin "got it right" in that regard. However, there is one significant aspect of the comic strip with which those with a biblical creationist view would take issue. The comic strip is based on the notion that those who cavorted with dinosaurs were Stone Age cavemen.

The Bible clearly states that God's six-day creation, including man and the animals, was very good (Genesis 1:31). There are a number of references in the Bible where men were described as living in caves, including even King David. But, there is no progressive evolving indicated in the Bible where apes became men over long eons of time. Adam was able right away to name all the land animals and birds (Genesis 2:20), an indication of exceptional intelligence. In Genesis 4:20–22 we learn that not long after the Fall of man into sin and death people were adept at agriculture, bronze and iron forging and the manufacture and playing of musical instruments. Noah's building of the ark required considerable architectural, construction and logistics skills.

After the Flood and after God scattered the people groups out from Babel, confusion was so rampant that many were reduced to using stone tools. This was a devolving of technology opposite in direction from the current secular view that man has always been evolving to be more and more intelligent and skilled. The pre-Flood people had great skills and abilities that were lost because of the destruction of the Flood (with only eight people left to transfer the technology of the past), and the language confusion at Babel. It has taken many centuries for much of that lost technology to be rediscovered and there are indications that even in the 21st century mankind has yet to relearn some of the technology known to pre-Flood ancients.

Question 10: Did men really ride dinosaurs like Alley Oop does in the comics?

Answer: The Bible does not mention anything one way or another about men riding dragons (dinosaurs). In the book of Job, Behemoth and Leviathan were described as wild beasts that would be difficult for anyone other than God to control. There are, however, a number of post-Flood documents and artifacts that indicate men did at times have very close contact with dinosaurs. The dinosaur fossils extracted from the rock record do not provide information that would help us know whether or not dinosaurs could be tamed or domesticated.

Dinosaurs and the Bible

Question 11: How accurate are the Alley Oop depictions of dinosaurs?

Answer: Taking the strip of Figure I–3, I can make the following general comments:

- Alley Oop's dinosaur, Dinny, was patterned by V.T. Hamlin after a *Brontosaurus*. That would make Dinny a created kind #11 or a Long-Necked Slender-Headed dinosaur. The dermal spines on Dinny can be considered accurate, as some discoveries have been made that indicate some Long-Necked dinosaurs did indeed have them. However, Dinny's teeth are not accurately depicted because kind #11 dinosaurs had pencil-like teeth in the front of their jaws and no teeth in the back. Another problem with the comic depiction is that Dinny's head is proportionally much too large compared to his body for a kind #11 dinosaur.
- The red-colored kind #15 Tyrant Bipedal dinosaur in the strip looks pretty realistic according to many current in-the-flesh reconstructions.
- The 19th and 20th century reconstructions of the largest dinosaurs always had them shown as sluggish beasts dragging their limp tails as they slowly trudged along. Today it is thought that they were much more agile and quick. However, it is unlikely in my opinion that the largest dinosaurs would have been able to do much jumping. They would probably move much like today's elephants and rhinos due to the physics of large size.
- Of course we do not know the color of the skin of dinosaurs so we do not know if the current Alley Oop cartoonists Jack and Carole Bender have them correctly colored or not. Thankfully, the Benders have not yet succumbed to evolutionist speculation that dinosaur skin was covered with feathers. That speculation is a result of so many secular paleontologists believing there must be a direct macroevolutionary connection between dinosaurs and today's birds.

Section II

Dinosaur Definition

What Are Dinosaurs?

DINOSAURS ARE TERRESTRIAL animals that have gone extinct as far as has been scientifically determined at the time of the writing of this guidebook. Of the fifteen North American created kinds identified in this guidebook, the only ones that I am aware of for which there is evidence that they may be still alive are the Long-Necked ones that have been reportedly seen by natives in deep Africa. Known by the locals as *Mokele-mbembe,* these creatures have been the object of a number of expeditions by cryptozoologists who would like to affirm their existence and then capture them alive if possible. I have not heard of any reports that dinosaurs have been spotted recently anywhere in North America.

The word dinosaur means "terrible lizard" in Greek, and up until the late twentieth century most paleontologists thought that dinosaurs were reptiles pretty much like those that exist today. Much evidence has been discovered that dinosaurs laid eggs like reptiles (see Section IV for more on this). There is disagreement today as to whether the dinosaurs were/are cold blooded (ectothermic) like the reptiles of today or warm blooded (endothermic) something like mammals. Creation scientists can accept that God could have decided to make the dinosaurs different from reptiles and other existing animals. In secular paleontology, where it is believed dinosaurs must have evolved from some other animal or animals over millions of years, there are competing theories for what the common ancestors were. Today, most secular paleontologists believe that dinosaurs evolved into birds.

The creationist hypothesis proposed in this guidebook is that God originally created these animals on day six according to kinds like the other animal kinds such as the cat, dog, horse and bear kinds. The fossil dinosaurs found in the rock record had developed genetic variation over the approximately 1,600 years since the creation and prior to the Flood when they were rapidly buried. The reason they had the pre-flood capacity for this variation was that God created them with the necessary information in their DNA to allow for adaptation to varying environmental and ecological conditions. When life-forms are unable to adapt quickly enough to changing conditions, they go extinct. However, I surmise that the major condition that forced these pre-flood changes in the animals was the *curse* instituted by God at the time of the Fall of mankind. This sin-caused Fall resulted in some of the dinosaurs becoming more and more evil, along with mankind, with some even changing from being plant eaters to meat eaters against the expressed instructions of God that are recorded in the Bible. There are a number of dinosaur fossils that indicate that animals ate each other during the pre-Flood times. The whole creation was affected by the sinful actions of man at the Garden of Eden and will continue to be so until the return of Christ when a new earth will be created.

While the dinosaurs were created with a fairly limited number of kinds and then developed considerable variation within those kinds prior to the judgment of the worldwide flood, the study of their fossils indicates that God created them with just two basic hip designs. So far, all dinosaurs that have been discovered in the sedimentary layers of the earth seem to be able to be placed in one of these two pelvic or hip-design categories. They had either what is called "lizard-hips" (aka saurischian) or "bird-hips" (aka ornithischian). Figure II–1 compares these two designs using sketches and actual replicated bones are shown for the two designs in Figures II–2 and II–3.

Both hip designs had three main parts that can be identified as similar. The top part is called the *ilium* and is more massive in the saurischian compared to the ornithischian. The forward part is called the *pubis* in both designs. The pubis bone points down and forward in the saurischian, but is swept back and lays along the backward-facing *ischium* bone in the ornithischian. The ischium in the saurischian design is separated from the pubis and is pointed back and down.

Dinosaur Definition

Fig. II–1: Lizard-Hip design (upper) and Bird-Hip design (lower)
—By M. Pike

The evolutionist must ponder how these two quite different hip designs could have ever evolved from some unknown ancestor. He must have faith that given enough time anything could have somehow happened.

The creation scientist, especially one with engineering experience, appreciates the fact that the hip structures are designed to provide strength and efficient function and that the thought that their existence is due to chance requires faith beyond the confines of logic. In observable reality all complex design results only from intelligent designers. Also, if we have these animals reconstructed properly, God showed an incomparable engineering skill by using both the saurischian and the ornithischian designs for both bipedal and quadrupedal dinosaurs (see Table II–1).

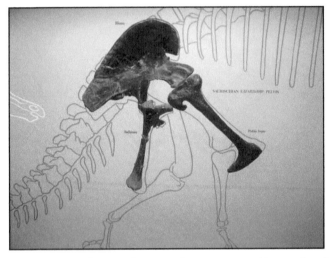

Fig. II–2: Lizard-Hipped Design for Dinosaurs (Saurischian)

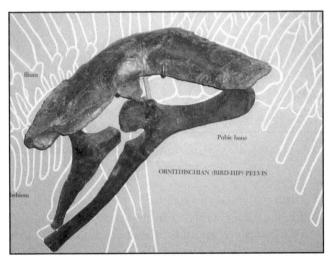

Fig. II–3: Bird-Hipped Design for Dinosaurs (Ornithischian)

Dinosaur Definition

In Section IX of this guidebook are details for all of the identified kinds of North American dinosaurs, and denoted for each kind are the type of hip design and stance that they had, and as also shown in the table below:

TABLE II–1

KIND NO.	KIND DESCRIPTION	HIP DESIGN	STANCE
1.	Armor-Backed	Bird-Hipped	Quadrupedal
2.	Bipedal Browser	Bird-Hipped	Bipedal
3.	Club-Tailed	Bird-Hipped	Quadrupedal
4.	Duck-Billed	Bird-Hipped	Quadrupedal
5.	Horn-Faced	Bird-Hipped	Quadrupedal
6.	Horn-Nosed Bipedal	Lizard-Hipped	Bipedal
7.	Killer-Clawed	Lizard-Hipped	Bipedal
8.	Lithe, Fast Running	Lizard-Hipped	Bipedal
9.	Long-Necked Big-Clawed	Lizard-Hipped	Bipedal
10.	Long-Necked Boxy-Headed	Lizard-Hipped	Quadrupedal
11.	Long-Necked Slender-Headed	Lizard-Hipped	Quadrupedal
12.	Ostrich-Like	Lizard-Hipped	Bipedal
13.	Plate-Backed	Bird-Hipped	Quadrupedal
14.	Thick-Headed	Bird-Hipped	Bipedal
15.	Tyrant Bipedal	Lizard-Hipped	Bipedal

All of these dinosaurs of either hip design were spoken into existence by God on day six of creation along with many other kinds of animals. Later on day six God created the first man (Adam) from the dust of the earth and the first woman (Eve) from a rib of Adam. Unlike the animals and every other life-form, Adam and Eve were uniquely made in the image of God.

Those many secular paleontologists who believe that birds evolved from dinosaurs think they evolved from the Lizard-Hipped category of dinosaurs rather than the Bird-Hipped category—just opposite from what common sense might dictate.

The often-used differentiation based on carnivorous (theropod) versus herbivorous (sauropod) types cannot be exactly correlated for the created-kinds taxonomy in this guidebook. While all of the Tyrant Bipedal kinds could possibly be placed in the theropod type grouping and the Long-Necked kinds could possibly be placed in the sauropod type grouping, it may not always work for all the other created kinds and so I don't emphasize herein the terms sauropod, theropod, ceratopsid, ornithopod etc. At the time of creation none of the dinosaurs were carnivorous and there evidently was no standard path of variation taken to carnivorous behavior during the pre-flood times. It seems that the only thing that can be determined for sure by a study of teeth and jaw structures is that not all dinosaurs became carnivorous prior to the Flood.

Dinosaur Skeletons

Anyone who studies dinosaur fossils and/or their reconstructions will want to understand the anatomical terminology and conventions used by paleontologists. Figure II–4 is a depiction of the major skeletal parts of a typical Long-Necked kind of dinosaur. The other kinds of dinosaurs would have the same basic skeletal structure terminology even if they varied in detail.

When looking at animal skeletons the conventions are that the "anterior" view is from the front of the animal, the "posterior" view is from the rear, the "dorsal view is from the top, and the "ventral" view is from the bottom. A "lateral" view is one looking at the side and a "medial" view would be looking toward the middle of the animal. The glossary also provides definitions and descriptions of various skeletal and anatomical words.

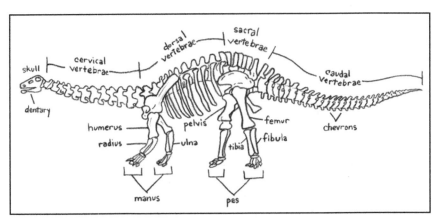

Fig. II–4: Some Basic Skeletal Parts of Dinosaurs —By M. Pike

The development of a biblical creationist worldview allows one to better appreciate the diversity that God has applied to the design of the skeletons of dinosaurs. An example is the skeletal part denoted as the "furcula" or wishbone. Until quite recently these furculas were reconstructed by paleontologists in various locations in the ribcage on the Tyrant Bipedal kind of dinosaurs. See Figures IX–15.48, IX–15.49, and IX–15.50 for the current assumed location for these relatively small boomerang-shaped bones; located across the chest. Flying birds have V-shaped furculas similarly located that assist in their being able to fly. Obviously, assistance for flying is not the function of the furcula in Tyrant Bipedal dinosaurs! Figure IX–15.52 is a photo of a full-size replica *Allosaurus* furcula.

Dinosaur Skulls

As is the case with most kinds of animals there usually is more variation in dinosaur skulls and teeth than there is in the skeletons. Carefully look at the following sketches of four different kinds of dinosaur skulls. Notice that even though they have similar bone parts to their skulls, the variation is considerable. Table II–1 is the key for the various skull bone abbreviations in the skull sketches. In a manner similar to the genus names of the dinosaurs the skeletal and skull bones are named in Latin or Greek. For examples, pes comes from *pes* which is Latin for "foot," nasal comes from *nas* which is Latin for "nose," maxilla comes from *maxilla* which is Latin for "jawbone," and rostral comes from *rostratus* which is Latin for "beaked."

For some kinds of dinosaurs the skulls are less likely to be found as fossils. Fossil parts for Long-Necked kinds that have heads that are small compared to their bodies are often found minus heads. On the other hand, the skulls for the Horn-Faced dinosaurs are much more common than are their skeletons because of the massive size of the heads compared to the bodies.

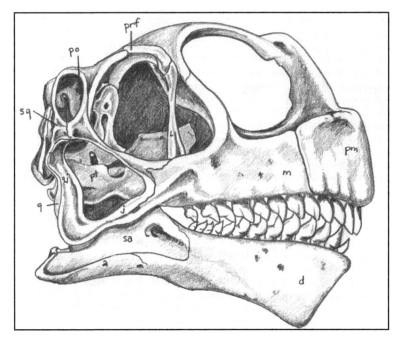

Fig. II–5: Skull of a Long-Necked Boxy-Headed Dinosaur (*Camarasaurus*)
—By M. Pike after Romer

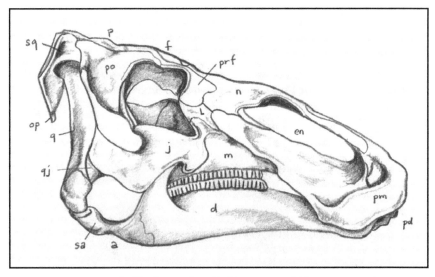

Fig. II–6: Skull of a Duck-Billed Dinosaur (*Edmontosaurus*)
—By M. Pike after Romer

Dinosaur Definition

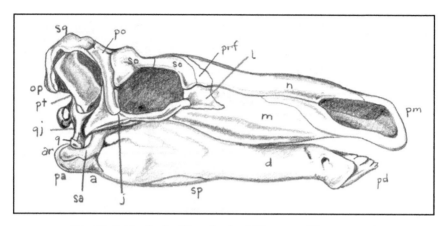

Fig. II–7: Skull of a Plate-Backed Dinosaur (*Stegosaurus*)
—By M. Pike after Romer

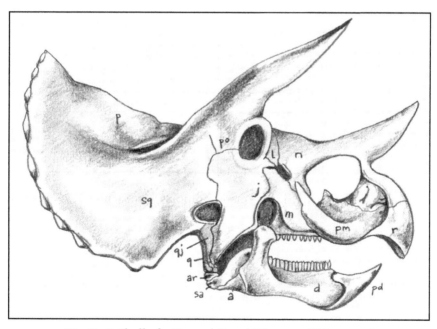

Fig. II–8: Skull of a Horned-Faced Dinosaur (*Triceratops*)
—By M. Pike after Romer

ABBREVIATION	WORD(S)
a	angular
ar	articular
d	dentary
en	external nares
f	frontal
j	jugal
l	lacrimal
m	maxilla
n	nasal
op	opisthotic
p	parietal
pd	predentary
pm	premaxilla
po	postorbital
prf	prefrontal
pt	pterygoid
q	quadrate
qj	quadratojugal
r	rostral
sa	surangular
so	supraorbital
sq	squamosal

Table II–1: Key to Skull Bone Terminology in Figures II–5 through II–8

Como Bluff—Lots of Bones!

One of the world's greatest and best known fossil beds of dinosaur remains is located at Como Bluff, Wyoming. Discovered and first worked by dinosaur hunters such as Marsh and Cope in the 1870s, genera such as *Allosaurus, Stegosaurus, Camarasaurus, Diplodocus* and *Brontosaurus* were discovered there. From several dig sites at this location many dinosaur, mammal, turtle, and other fossils were shipped to the Smithsonian, American, and Peabody museums for study, reconstruction and display throughout the latter third of the 19th century.

Fig. II–9: Old Aerial Photo of the Como Bluffs

After the museum-sponsored fossil hunters had mostly left the area for new digs there were so many discarded fossilized bones left lying around that an enterprising individual by the name of Thomas Boylan utilized them to make a fossil cabin museum in 1933. He also constructed a large stone house about the same height and length as a *Brontosaurus* dinosaur next to the fossil cabin. At the time of the writing of this guidebook, the property was closed awaiting a buyer. The fossil cabin and the adjacent house are located eight miles east of Medicine Bow, Wyoming, on Highway 30.

At the front entry to the house are two large bones embedded in the wall that look to be leg bones from a large dinosaur. The bones would be prize possessions for many natural history museums. A sign over the sealed entry door to the fossil cabin museum reads that it is made up of 5,796 individual dinosaur bones. As a dinosaur bone collector, it would be the author's opinion that the bones making up the cabin could be sold for many hundreds of thousands of dollars at rock shops if they were currently available for sale as individual pieces.

Fig. II–10: Fossil Cabin Museum near Como Bluff in Wyoming

Fig. II–11: Boylan's House next to the Fossil Cabin near Como Bluff

Dinosaur Definition 25

Fig. II–12: Como Bluffs House with Large Leg Bones at Entry Door

Fig. II–13: Close-up Photo of Dinosaur Bones in Fossil Cabin Museum

The vast assortment and quantity of dinosaur and other animal bones at Como Bluff, Dinosaur National Monument, and other similar dinosaur graveyards throughout the world can be best interpreted to be the result of the cataclysm described in chapters 7–9 of Genesis.

The condition of the bones and skulls described in this section of the guidebook remind us that in most cases there is not much left of them from God's judgment of sin some 4,500 years ago. The condition and completeness of the remnants from that judgment also make it difficult to really know what these animals looked like. Therefore, those who study them, whether they are secular or biblical paleontologists, often end up cloaking their descriptions and reconstructions with large amounts of speculation.

Section III

Dinosaur Footprints

Ichnology

ICHNOLOGY IS A branch of paleontology devoted to the study of individual fossil footprints and sequences of footprints called trackways. These footprints comprise the most common trace fossil and in places are much more prevalent than the bones of the animals that made them. It is estimated that there are billions of dinosaur tracks alone in the rock record.

Fig. III–1: Animal Footprints Preserved in Rock

No dinosaur has yet been discovered standing in its trackway so there is no sure-fire method to identify track makers from their tracks. Informed guesses can be made based upon differences in the shape, number of toes, length, width and depth of the tracks. However, as ichnology research advances, the information gained can contribute more and more to our knowledge of the animals that left the footprints. A taxonomic system (ichnotaxonomy) is being developed for the tracks where scientific names are assigned to the footprints themselves. Sometimes it is as simple as ending existing scientific animal names in "pes," "pod," "podus," or "ichnus." For example, the footprint thought to be from a genus *Brontosaurus* could be named *Brontopodus*. In most cases it has not been possible to positively identify a dinosaur from its track.

Fig. III–2: *Grallator* Dinosaur Footprint

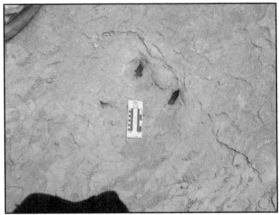

Fig. III–3: Large Three-Toed Dinosaur Footprint near Moab, Utah

There is a fairly common three-toed dinosaur footprint that has not been identified at the genus level and is named *Grallator*, with no direct name-connection to any particular dinosaur.

Fig. III–4: Large Dinosaur Footprint in Coal from Price, Utah Area

Sites with Many Tracks

In many places there have been discovered what are called "megatracksites" where individual sites have large numbers of tracks and trackways. In the United States these large sites are located in what looks like a wide path from Texas up northwesterly to Utah, Colorado, and Wyoming. Perhaps the most famous megatracksite is located in central Texas in the Glen Rose area where many dinosaur tracks have been under study for almost a century. The Purgatoire River Trackway in southeast Colorado has nearly one hundred separate trackways that have been mapped so far. Other significant North American megatracksites are located in New Mexico, Utah, Wyoming, and even in Connecticut.

Fig. III–5: Iconic Photo of Sauropod Footprint at Paluxy River near Glen Rose, Texas

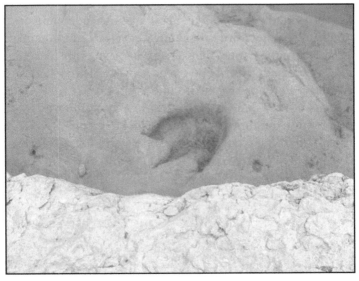

Fig. III–6: Author's Photo of Three-toed Dinosaur Footprint at the Blue Hole, Dinosaur Valley State Park

How Did the Tracks Form?

Our understanding of the Flood conjures up visions of great catastrophe, and so how could these dinosaur footprints be preserved? Wouldn't any footprints placed during the Flood be wiped out by the actions of the Flood?

The first thing that needs to be remembered is that the Genesis flood lasted for a year. We can expect that whatever the flood mechanisms were, they and the geologic results at any one location would not have been identical for each and every hour of the Flood. A lot of things can happen in a year.

Here follows a list of logical requirements for the preservation of dinosaur footprints:

- The dinosaurs would have had to have stepped into a material like mud to make the imprints in the first place. (The worldwide flood definitely had a lot of water-laden sediments, so that matches.)
- The imprints could not have been left as originally made for very long or they would have been destroyed by something. So, the imprints had to have been fairly quickly buried just like all fossils. (A concept of tidal influences and/or periodic water waves could be understood here.)
- The burial sediments had to be of a type that would preserve the track and not destroy them. (The rock record shows evidence of large areas of one kind of water-borne sediment as well as many areas where multiple kinds of sediments are mixed together.)
- The buried sediments must have hardened quickly or the tracks would have been destroyed by worms, animals and who knows what. (Everything in the fossil record has been preserved because of the hardening and cementing of organic and inorganic materials.)

There is little consensus on many topics in the field of ichnology, and numerous secular and creationist scientists are studying these tracks and trying to understand the processes that made them. Their interpretations directly reflect their presuppositions and so creationists develop hypotheses that match the biblical record while secularists develop hypotheses that match evolution and millions of years. *Dinosaur Challenges and Mysteries* is an excellent book by creationist Michael Oard that is largely devoted to this subject.

Fig. III–7: Dinosaur Trackway along Paluxy River Being Excavated by Roland T. Bird

Are Human & Dinosaur Footprints Together?

Fig. III–8: Footprints ©2008 Brent Giles and Paul Abramson

The biblical creationist understands that the Bible is perfectly clear that dinosaurs and man were created together on day six. Therefore, man and dinosaurs lived together. However, the direct evidence for this position from the rock record is not quite so clear.

Evolutionists "know" that there can be no physical evidence that humans and dinosaurs lived together because their faith-driven long-ages paradigm does not allow for it. Therefore, no matter how clear any physical evidence might be for humans and dinosaurs living together, evolutionists must immediately reject it.

These diametrically opposite perspectives on human and dinosaur footprints together cannot be more clearly seen than at the Paluxy River in Glen Rose, Texas. As is the norm in the United States, the "official" interpretation of the tracks at Dinosaur Valley State Park where these trackways are located is based on atheistic secular presuppositions of evolution and millions of years. Evolutionists date the tracks at over 100 million years old. Biblical creationists believe they are a result of the worldwide flood about 4,500 years ago.

History of the Paluxy River Tracks

The dinosaur tracks were first discovered in 1908 as a result of a large Paluxy River flood that year that wiped away large amounts of limestone. Initially the three-toed tracks that were uncovered and then discovered as a result of the flood were popularly thought to be from giant turkeys.

By 1932 those tracks and other later-discovered five-toed tracks were properly identified as from dinosaurs. Professional paleontologist Roland T. Bird from the American Museum of Natural History began studying the tracks in 1938. Within a few years, Bird was able, with great effort and the help of WPA workers, to excavate a fairly large section of one Paluxy River dinosaur trackway (See Figure III–7). Eventually he was able to reassemble the trackway at the American Museum in New York where it remains displayed beneath a reconstructed dinosaur skeleton. *National Geographic* magazine coverage in the early 1950s was instrumental in bringing Paluxy dinosaur tracks to worldwide attention.

After the release of the book *The Genesis Flood* in 1961 by John Whitcomb and Henry Morris, creationists took interest in the tracks and identified what they thought were human footprints at several locations along the Paluxy. That ignited an interpretation battle that has continued unabated ever since. The heated arguments are not about the dinosaur footprints, but the human ones. There are a number of evidences that have been used by those believing that the Paluxy River rock layers include dinosaur and human footprints together. Two of them are the Taylor Trail and the Delk Print.

The Taylor Trail

Mr. Stan Taylor excavated a dinosaur trackway and an associated human trackway along the Paluxy from 1969 through 1972. The excavated result is shown in Figure III–9.

Fig. III–9: Taylor Trail of Dinosaur and Human Footprints along the Paluxy River in Texas

What many creationists believe was found are fourteen human tracks at the same rock layer level as a trail of three-toed dinosaur tracks. The line of dinosaur tracks are similar to many others excavated along the Paluxy over the years and are oriented at about thirty degrees to the line of what look to be human footprints.

The fourteen footprints are consistent in length and were laid down in an unbroken left-right pattern. Half of the fourteen prints had what looked to be individual toes. In Figure III–10 is an image of the print identified as "-3B."

Dinosaur Footprints

Fig. III–10: Human Footprint -3B from the Taylor Trail

The Taylor Trail footprints are consistent with the unique form and construction of the human foot and are spaced at an expected length for the human gait. Secular anti-creationists have developed theories for why the footprints are really from dinosaurs and only look human. Some creationists also remain skeptical of their authenticity because some footprints that initially looked human eroded to not look human over the years after discovery.

The Alvis Delk Print

The fossil footprint known as "The Alvis Delk Print" was discovered by Mr. Alvis Delk. It was found as a loose slab along the Paluxy River about a mile north of Dinosaur Valley State Park. Upon turning the slab over Mr. Delk noticed an imprint of a three-toed dinosaur and took the slab home to add to his large fossil collection.

Eight years later Mr. Delk decided to clean off the dried mud that covered the print side of the slab. The cleaning resulted in his discovering the view shown in Figure III-11, which looks to be a human footprint intruded upon by the dinosaur print.

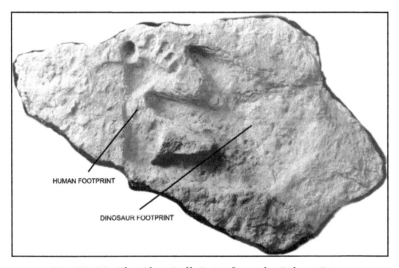

Fig. III–11: The Alvis Delk Print from the Paluxy River

Eight hundred spiral CT scans of the fossil slab have been undertaken that indicate it is authentic and not a carving. The Delk fossil, along with other Paluxy River footprint fossils, are located for viewing at Dr. Carl Baugh's Creation Evidence museum located next to the river in Glen Rose, Texas.

Conclusion

People holding to evolutionary presuppositions will never be convinced that men and dinosaurs co-existed, no matter the evidence. On the other hand, biblical creationists have long been open to the possibility that the rock record would eventually reveal direct scientific evidences that the two were contemporaneous. Others who come to believe that the evidence from the rock record shows that human and dinosaur footprints were laid down at the same time in history logically should come to understand that the evolutionary paradigm is false.

Section IV

Dinosaur Eggs

Eggs from Dinosaurs?

IT SEEMS PRETTY clear from the fossil evidence that dinosaurs reproduced by the production of hard-shelled eggs. Since fossilized eggs have been discovered in the sedimentary layers in which dinosaur bones have been found, it follows that the one is probably related to the other. The creation scientist is more cautious about this inference since the effects of the Flood could have caused considerable separation of the eggs from the bones at times and at places.

Most of the dinosaur egg material found so far is in the form of shell fragments, although quite a number of nearly complete fossil eggs have been found too. If fragments or an egg or two are all that are in hand

Fig. IV–1: Long-Necked Dinosaur Egg Shell Fragment
About 3/16" (5 mm) Thick

there is no way to know for sure that any particular egg corresponds to any particular dinosaur kind. Sometimes the eggs have been found in "nests" or "clutches," but the eggs and the nests show a large amount of diversity of form.

At one time secular paleontologists, for the most part, believed that dinosaur eggs were positive evidence that dinosaurs were reptilian. That is no longer the case, since most secularists of today believe that birds are dinosaurs and therefore there should be a close relationship between bird eggs and dinosaur eggs.

Personally, the more I study dinosaurs the more I think they were not reptiles, birds or mammals but rather ... dinosaurs. God made them as different from reptiles, birds and mammals as those animals are different from each other.

Dinosaur Embryos

Embryos have been discovered within the eggs for a few different kinds of dinosaurs. In those extremely rare occurrences, information can be gained that helps connect dinosaur kinds to the size, shape and construction of the eggs. However, the current hypotheses developed from dinosaur embryo evidence far outweigh definitive conclusions. A complication in the attempt to connect egg to dinosaur is that the dinosaur embryos so far found

Fig. IV–2: Photo of a Dinosaur Egg with Embryo (Museum of Rockies)

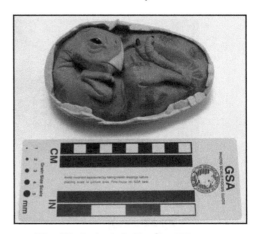

Fig. IV–3: Artist's Replica Dinosaur Embryo to Scale

Dinosaur Eggs 39

are so similar it is very difficult to use embryos to assign defining characteristics for the different kinds.

Some of the dinosaur eggs with embryos that have been found were located in Montana, and Figures IV–2 and IV–4 are from the Museum of the Rockies in Bozeman, MT.

Fig. IV–4: Reconstructed Dinosaur Egg and Embryo
(Museum of the Rockies)

What Dinosaur Eggs Look Like

Even with all of the confusion and uncertainty, paleontologists have in a number of cases applied identification to dinosaur eggs according to the animals they believe laid them. The egg photos that follow are labeled according to dinosaur kind as identified by the museum that displayed them. The eggs in Figures IV–5 through IV–12 have been represented to be actual eggs, not replicas by the museums. I think the identifications for these eggs should be considered tentative at best. Most of the dinosaur eggs I have seen range from six inches to ten inches in overall size. Figure IV–13 shows the sizes of two of my personal dinosaur egg replicas.

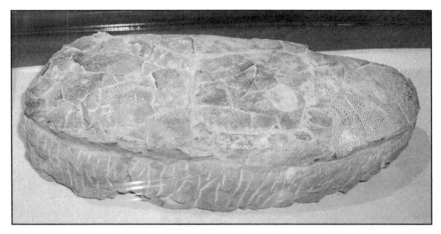

Fig. IV–5: Tyrant Bipedal Fossil Dinosaur Egg
(Creation Museum, Kentucky)

Fig. IV–6: Long-Necked Fossil Dinosaur Egg
(Creation Museum, Kentucky)

Dinosaur Eggs 41

Fig. IV–7: Tyrant Bipedal Fossil Dinosaur Egg
(Creation Museum, Kentucky)

Fig. IV–8: Duck-Billed Fossil Dinosaur Eggs
(Creation Museum, Kentucky)

Fig. IV-9: Killer-Clawed Bipedal Fossil Dinosaur Egg Cluster
(Museum of the Rockies, Montana)

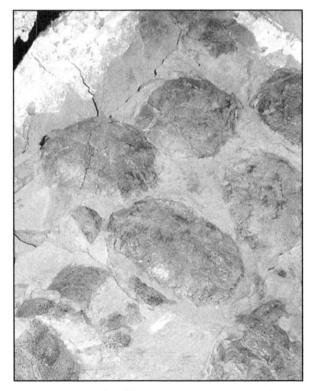

Fig. IV-10: Duck-Billed Fossil Dinosaur Egg Cluster
(Museum of the Rockies, Montana)

Dinosaur Eggs

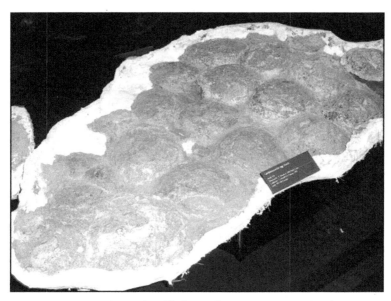

Fig. IV–11: Duck-Billed Fossil Dinosaur Egg Clutch
(Museum of the Rockies, Montana)

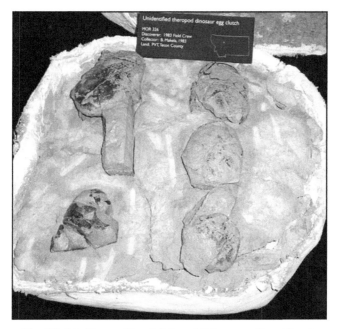

Fig. IV–12: Tyrant Bipedal Fossil Dinosaur Egg Clutch
(Museum of the Rockies, Montana)

Fig. IV-13: Full Size Replicas of Fossil Dinosaur Eggs

From the previous photos it can be seen that most of the identified dinosaur eggs are either mostly oblong in shape or of a typical spherical shape. The surfaces of the eggs seem to usually be fairly smooth, but some eggs exhibit some exterior textural patterns. As with the skin of dinosaurs, no way has yet been developed to determine the original colors of the eggs.

Are Dinosaur Eggs Available to the Public?

Complete fossilized dinosaur eggs are rare enough that authentic single specimens are quite expensive to purchase. Good quality fossil eggs start at about five hundred dollars each and fossil egg clutches normally sell for many thousands of dollars. Replicas are readily available for one tenth or less the prices of actual fossil eggs. Most authentic dinosaur eggs available for sale in North America originate in Asia, especially China, with North America specimens exceedingly rare and expensive.

On the other hand, dinosaur egg shell fragments are not difficult to find and are available for sale from many retail fossil suppliers. One-inch square dinosaur egg shell fragments can often be purchased for about ten dollars each.

Section V

Dinosaur Death

Extinction Theories

UNTIL THE LAST two decades of the twentieth century secular scientists in general, and paleontologists specifically, had not taken the question of the reason for the extinction of the dinosaurs very seriously. While it was generally accepted in the secular world that all dinosaurs became extinct about 65 million years ago, the general public was allowed to ponder a plethora of secular hypotheses for the supposed extinction, including some that were humorous or even wacky. A few of the weirder hypotheses commonly discussed in popular science explanations were:

- Mammals ate all the dinosaur eggs before they could hatch.
- Too many males or females were born to allow propagation.
- The dinosaurs developed extreme allergies.
- The dinosaurs died of constipation.
- Dinosaur eggshells became structurally too weak.
- Dinosaurs ate foods that poisoned them.
- Dinosaurs could not compete for food with mammals.

Since none of these (and a number of other similar ideas) was particularly based on any real science, and they of course had no way to be tested, it would be fair to relegate them all to the genre of science fiction. Paleontology is a field of study that is mostly historical and there is not much that can be scientifically tested, but these extinction hypotheses provide a good idea of the latitude allowed the secularists in their ongoing speculations concerning origins issues.

By the start of the 21st century, however, the secular debate for the cause of the dinosaur extinction had settled to just a couple of main hypotheses, both assuming mass extinction and based on the discovery of a thin layer of the element Iridium near the so-called K/T boundary in Europe. The two hypotheses being debated today are the (comet/meteorite/asteroid) impact theory and the volcanism theory. Thousands of papers and books have already been published on this contentious topic, but it is not in the scope of this guidebook to explain either of these atheistic conjectures to any degree. No, the emphasis of the rest of this section will be on a better dinosaur extinction explanation that is *never* touched on in secular debates—the worldwide flood described in the book of Genesis in the Bible.

Rapid Burial

Most of what is known about dinosaurs has been gleaned from their fossils found in the sediments that make up the rock record. Secular popular fossilization explanations notwithstanding, all knowledgeable secular or creationist experts agree that the natural formation of fossils requires rapid burial in waterborne sediments. Dinosaurs were not likely ever turned into fossils by the gradual covering of their dead bodies by the normal operation of rivers, streams, and lakes as is usually illustrated in secular dinosaur guidebooks. For an entire dinosaur skeleton to have been fossilized it would have required it to have been buried rapidly and completely with massive amounts of water-borne sediments like those we can imagine were a part of the worldwide flood at the time of Noah.

Graveyards

Dinosaur fossils are often found in jumbled masses, bone beds, or graveyards. Sometimes the bones are partially articulated but usually they are totally disarticulated. In some locations the bones are all of one kind or just a few kinds. In many locations the bones are a mixture of numerous kinds of dinosaurs as well as other life-forms. A complete fossil skeletal find is extremely rare. Part of the task of the paleontologist is to carefully document each fragment of these bone beds and keep track of where the bones were found relative to each other. The condition of the fossils can be taken to indicate that the imagined drastic actions of the

Dinosaur Death

worldwide flood were the cause. The result of catastrophe is what we most often see at the site of dinosaur fossil finds, not gradual uniformitarian geological action.

Fig. V–1: Dinosaur Bone Bed Excavation in Utah
(Dinosaur Bones Have Been Painted Black)

Taphonomy

Taphonomy is the technical term paleontologists use for their study of the conditions and processes by which organisms have been fossilized and preserved in the rock record.

After the actions of the Flood killed the dinosaurs, their dismembered remains were likely water-transported to a location some distance away from where they originated. Those that were fossilized and later discovered ended up at some specific place in the hardened sediments of the catastrophe. Then through the work of erosion and earth movements the remains were later found on or near the surface. In most cases only the hard parts like bones, skulls, teeth, and claws were preserved, but sometimes dinosaur softer parts are found as imprints, molds or mummified remnants. The bones are sometimes found pretty much in the same physical condition they were in when the dinosaur died about 4,500 years ago—that is as normal bone. Oftentimes though, the cellular structure of the original bone is replaced by mineral substances such as silica. It is thought that the silica or other mineral was in ground water moving through the sediments, and the carbon in the bone was replaced molecule by molecule until the bone became hard like a rock. This process is called permineralization.

Most fossil bone will fairly rapidly disintegrate from the actions of the environment once it is released from its location in the sediments. Fossil hunters go to great effort to protect their finds from detrimental environmental effect so they can be studied and described back at the laboratory. That means that they often have to use a lot of glue to keep the bones from falling apart as they are recovered, transported, and studied.

Waterborne sediments can harden to rock in a year or so, and it does not take millions of years for fossils to develop either. However, there is little evidence that fossils are being made in the geologic conditions we are experiencing today. I think it is clear that most of the fossils (especially dinosaur fossils) discovered required the special cataclysmic conditions inherent in the Flood. These conditions were not only special in severity but also special in the fact that they occurred over a one-year period only one time in history.

Dead Dinosaur Posture

When fairly complete dinosaur skeletal remnants are found they very often are in what is called the "opisthotonic posture." These skeletons are in a characteristic pose where the head is thrown back, the hind limbs bent, and the tail extended.

Veterinarians have reported that animals often go into the opisthotonic posture shortly before they die, possibly due to muscle spasms. It is thought that the spasms may be the result of a central nervous system failure as the animals are dying. Dinosaur skeletons recovered in the death pose are shown in Figures V–2 and V–3.

Dinosaur Death

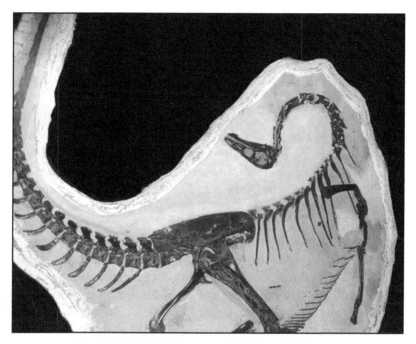

Fig. V–2: Typical Dinosaur Opisthotonic Posture

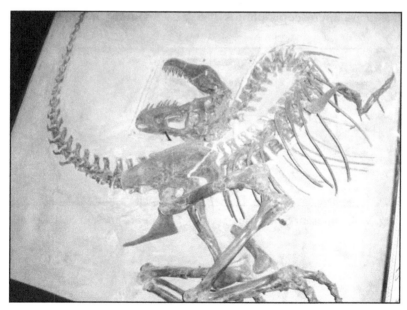

Fig. V–3: Dinosaur Pose as a Result of the Conditions of the Worldwide Flood?

This explanation fits well with the biblical explanation that everything in the Flood that had the breath of life in its nostrils died. What we see in the dinosaur death posture is a snapshot of the last dreadful moments of these creatures as they fought to stay alive.

Dinosaur Soft Tissue

Many discoveries all over the world have reported the existence of soft tissues inside fossils including dinosaur bones. Reliable scientific techniques have found soft materials such as chitin, elastin, fibrin, osteocalcin, keratin, hemoglobin, and collagen in various supposedly millions-of-years-old fossils. Even DNA is thought to have been found in dinosaur bone!

As the years since these discoveries turn into decades secular science continues to struggle to explain these soft-tissue findings within the evolutionary paradigm. Since they "know" that the tissue has to be at least 65 million years old, they dream of fossilization processes that could keep the dinosaur tissues that long. But, common sense tells us that these bones are thousands of years old, not millions. The findings go far to confirm the biblical creationist view that the dinosaurs in the rock record died in the Flood just thousands of years ago. Figures V–4 and V–5 are microscopic photos of soft tissue found in dinosaur bones.

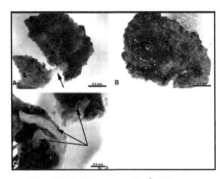
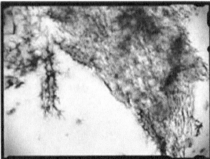

Fig. V–4: Dinosaur Soft Tissue   **Fig. V–5**: Dinosaur Soft Tissue

Section VI

Dinosaur Museums

Where to See the Reconstructions

THE FOLLOWING LIST is of North American museums that have significant permanent full-size skeletal dinosaur displays. A short description of some of the displays (according to genus name) is included for each. This is a list of the best museums so far as skeletal reconstructions that the author is aware of—there may be others. An attempt has also been made to list museums from a broad area of North America, so there is great disparity among them for quantity and quality. Museums modify their permanent displays on occasion, so it will behoove the visitor to make contact with the museum prior to traveling far to look at skeletons. All of these museums interpret their displays according to evolution and millions of years, excepting for the Creation Museum and the Glendive Dinosaur & Fossil Museum. Check the museum internet websites for exact locations, hours, admission fees and other current details.

CANADA

Alberta

1. **Royal Tyrrell Museum of Paleontology** in Drumheller:
   This is the greatest museum in Alberta and very likely the finest dinosaur museum in Canada. Skeletal reconstructions of the following dinosaurs are displayed:
   Euoplocephalus, Lambeosaurus, Edmontosaurus, Sauronitholestes, Ornithomimus, Tyrannosaurus, Gorgosaurus, Chasmosaurus, Albertosaurus, Prosaurolophus, Triceratops, and *Ornitholestes.*

Fig. VI–1: Royal Tyrrell Museum of Paleontology

2. **University of Alberta** in Edmonton:
   Mounted skeletons of *Stegoceras, Gorgosaurus,* and *Parasaurolophus* are on display in the Geology Department museum here.

Ontario

1. **Canadian Museum of Nature** in Ottawa:
   Collections by the Sternbergs and Canadian fossil collectors are the emphasis at this museum. The holotype for *Daspletosaurus* is on display.

2. **Royal Ontario Museum** in Toronto:
   Several original skeletons are on display here that were collected by Levi Sternberg. Skeletal reconstructions include *Tyrannosaurus, Stegosaurus, Triceratops, Barosaurus, Corythosaurus,* and *Parasaurolophus.*

UNITED STATES OF AMERICA

Arizona

1. **Museum of Northern Arizona** in Flagstaff:
   This museum displays a mounted *Dilophosaurus* and several skulls of *Coelophysis*.

California

1. **Natural History Museum of Los Angeles County** in Los Angeles:
   In the Dinosaur Hall of this museum can be seen skeletal reconstructions of *Tyrannosaurus*, *Triceratops*, *Stegosaurus*, *Allosaurus*, and *Camptosaurus*.

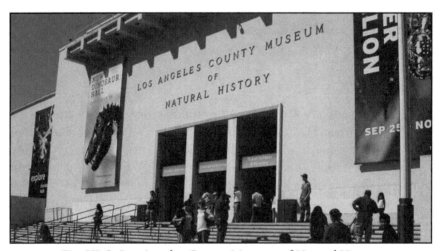

Fig. VI–2: Los Angeles County Museum of Natural History

Colorado

1. **Denver Museum of Nature and Science** in Denver:
   Skeletal reconstructions of *Tyrannosaurus*, *Allosaurus*, *Stegosaurus*, *Othnielia*, and *Diplodocus* are on display here.

2. **Dinosaur National Monument** in Dinosaur:
   In the Quarry Exhibit Hall are a few dinosaur specimens including a mounted *Allosaurus*. However, the biggest attraction is the world-famous Carnegie Dinosaur rock wall where the bones of

Allosaurus, Apatosaurus, Camarasaurus, Diplodocus, Stegosaurus and others are displayed as partially excavated from the sandstone.

Fig. VI–3: Entrance to Dinosaur National Monument

Connecticut

1. **Yale Peabody Museum of Natural History** in New Haven:
   The Great Hall of Dinosaurs exhibits skeletal reconstructions of *Apatosaurus, Camarasaurus, Stegosaurus, Camptosaurus,* and *Deinonychus*. Also found here is Rudolph F. Zallinger's famous "The Age of Reptiles" mural.

Illinois

1. **Field Museum of Natural History** in Chicago:
   There are a number of excellent skeletal reconstructions on display at the Field Museum including *Brachiosaurus, Deinonychus, Daspletosaurus, Parasaurolophus,* and *Stegosaurus*. The museum with the help of organizations like McDonalds and Walt Disney World paid $8.4 million for the fossil remains of the largest and most complete *Tyrannosaurus* skeleton found so far. Named Sue, this dinosaur skeletal reconstruction is a centerpiece for the museum.

Kentucky

1. **Creation Museum** in Petersburg.

    This quality museum has a full-size *Triceratops* skeletal reconstruction and a number of other dinosaur oriented displays. The museum is listed here not for the displays, but because it is one of the few that interpret the dinosaurs biblically.

Fig. VI–4: Answers in Genesis Creation Museum

Montana

1. **Glendive Dinosaur and Fossil Museum** in Glendive:

    This museum presents its displays in the context of a biblical creationist perspective. It is the largest dinosaur museum in North America that interprets the rock record in this manner and has skeletal reconstructions of *Acrocanthosaurus, Dromaeosaurus, Tyrannosaurus, Triceratops, Stegosaurus,* and *Pachycephalosaurus.*

2. **Museum of the Rockies** in Bozeman:
   The Museum of the Rockies is world-renowned for its large collection of dinosaur fossils. However, the dinosaur exhibit hall emphasizes *Tyrannosaurus* and *Triceratops* fossils with an impressive collection of skulls for these on display. There is a full-size bronze cast of a *Tyrannosaurus* skeleton outside the museum entrance.

Fig. VI–5: Museum of the Rockies

New York

1. **American Museum of Natural History** in New York City:
   This huge museum has a Hall of Ornithischian Dinosaurs and a Hall of Saurischian Dinosaurs. In the Ornithischian hall are reconstructions of *Corythosaurus, Anatotitan, Stegosaurus*, and *Triceratops*. The Saurischian hall features reconstructions of *Allosaurus, Apatosaurus, Deinonychus, Barosaurus, Coelophysis*, and *Tyrannosaurus*.

Ohio

1. **Cleveland Museum of Natural History** in Cleveland:
   Skeletal dinosaur reconstructions in this museum are *Tyrannosaurus*, *Triceratops*, *Coelophysis*, and *Allosaurus*. The holotype skull for *Nanotyrannus* can also be seen at this museum.

Oklahoma

1. **Sam Noble Oklahoma Museum of Natural History** in Norman:
   Found here are skeletal reconstructions of *Apatosaurus*, *Saurophaganax*, *Tenontosaurus*, *Deinonychus*, and *Pentaceratops*.

Pennsylvania

1. **The Academy of Natural Sciences of Drexel University** in Philadelphia:
   The 200-year-old museum of this institution has dinosaur skeletal reconstructions on display of *Tyrannosaurus*, *Avaceratops*, *Chasmosaurus*, *Corythosaurus*, *Deinonychus*, *Pachycephalosaurus*, and *Tenontosaurus*.

2. **Carnegie Museum of Natural History** in Pittsburgh:
   In 2005 the Carnegie Museum closed their old dinosaur hall, dismantled, and then reconstructed, their skeletal displays of *Diplodocus*, *Apatosaurus*, *Allosaurus*, *Stegosaurus*, and *Tyrannosaurus* to postures more in line with current thought.

Fig. VI–6: Carnegie Museum of Natural History, Pittsburgh

South Dakota

1. **Black Hills Institute Museum** in Hill City:
   This relatively small independent museum has a most impressive display of Tyrant Bipedal and other dinosaur skulls. Skeletal reconstructions that can be seen are *Albertosaurus, Tyrannosaurus, Triceratops, Edmontosaurus, Thescelosaurus*, and *Struthiomimus*.

Fig. VI–7: Black Hills Institute Museum, South Dakota

Texas

1. **Fort Worth Museum of Science and History** in Fort Worth:
   At this museum can be found information on the State Dinosaur of Texas named *Paluxysaurus* and a skeleton of *Tenontosaurus*.

Dinosaur Museums 59

Fig. VI–8: Fort Worth Museum of Science and History, Texas

2. **Houston Museum of Natural Science** in Houston:
   There are a large number of skeletal mounts here including *Tyrannosaurus, Stegosaurus, Gorgosaurus, Diplodocus, Deinonychus, Acrocanthosaurus, Edmontosaurus,* and *Triceratops*.

Fig. VI–9: Houston Museum of Natural Science, Texas

Utah

1. **Brigham Young University Museum of Paleontology** in Provo: This museum has a large number of dinosaur displays for the relatively small available space. Included are skeletal reconstructions of *Allosaurus, Torvosaurus, Camptosaurus, Gastonia,* and an unnamed long-necked kind.

Fig. VI–10: Brigham Young University Museum of Paleontology

2. **Cleveland-Lloyd Dinosaur Quarry** near Price: There is a visitor center at this BLM-operated quarry site that has several interesting dinosaur fossil displays about the bone graveyard here. Besides an *Allosaurus* skeletal reconstruction there are opportunities to see active quarry operations in separate buildings and take a 1 ½ mile moderate walk.

Dinosaur Museums

Fig. VI–11: Cleveland-Lloyd Dinosaur Quarry

3. **College of Eastern Utah Prehistoric Museum** in Price:
   In this museum are fine skeletal reconstructions of *Allosaurus, Camptosaurus, Chasmosaurus, Prosaurolophus, Animantarx, Utahraptor,* and *Gastonia*. It has numerous other interesting dinosaur displays as well.

Fig. VI–12: College of Eastern Utah Prehistoric Museum

4. **Dinosaur National Monument** in Jensen:
   See the entry for Dinosaur National Monument under Colorado.

5. **Utah Field House of Natural History State Park Museum** in Vernal:
   This museum displays dinosaur skeletal reconstructions of *Allosaurus*, *Stegosaurus*, and *Diplodocus* along with other associated dinosaur bones.

Fig. VI–13: Utah Field House of Natural History State Park Museum

Washington, D.C.

1. **National Museum of Natural History, Smithsonian Institution:**
   Skeletal dinosaur reconstructions of all the 'favorites' like *Triceratops*, *Diplodocus*, *Tyrannosaurus*, *Stegosaurus*, and *Allosaurus* are on display here.

Dinosaur Museums

Fig. VI–14: National Museum of Natural History, Smithsonian Institution

Wyoming

1. **Wyoming Dinosaur Center** in Thermopolis:
   This large dinosaur museum has skeletal reconstructions of *Bambiraptor, Allosaurus, Mymoorapelta, Othnielia, Supersaurus, Triceratops, Tyrannosaurus, Zuniceratops, Albertaceratops, Gastonia,* and *Maiasaura*; it also is the home for the Thermopolis specimen of *Archaeopteryx* from Germany.

Fig. VI–15: Wyoming Dinosaur Center

Section VII

Dinosaur Hunting

Who Digs for Dinosaurs?

AS MENTIONED IN the Introduction, most of the dinosaur fossils that have been dug up and scientifically studied are under the control of secular organizations. Historically, professional fossil hunters and paleontologists have been responsible for the discovery of most of these fossils from North America. Until the late twentieth century, by and large, amateurs were not particularly welcome at organized dinosaur fossil digs.

Times are changing and more and more creationist-oriented digs and museums are being organized. As this has transpired the secular organizations have also solicited more and more help from amateurs in their dino digs. Amateurs can move a lot of dirt at a much lower cost than can the professionals, and there are numerous secular museums and other organizations that sponsor amateur dino digs of some sort in North America.

An internet search is a good way to find the current organizations that sponsor these digs, whether secular or creationist. At the time of this writing, the two most active creationist groups that sponsor dino digs are the Foundation Advancing Creation Truth located out of the Glendive Dinosaur and Fossil Museum in Montana and Answers in Genesis in Kentucky. Most dino digs that are open to amateurs take place in Colorado, Montana, Wyoming, South Dakota, and Alberta. The secular dino dig operations will interpret the bones recovered according to evolution and millions of years. The creationist organizations will interpret them based on the Bible and the worldwide Flood at the time of Noah. Otherwise, the dino-dig groups have variable goals, terms, conditions, and costs for those amateurs that wish to join in. Some digs allow the diggers to keep some of the fossils, while others keep all fossils discovered.

Fig. VII–1: Glendive Dinosaur and Fossil Museum in Glendive, Montana

Collecting Dinosaur Bones

Most readers have probably heard of the "bone wars" of the late 19th century carried on between O.C. Marsh and E.D. Cope. That particular exuberant competition for fossils finally came to a close upon the death of the two main protagonists. In the late twentieth century another bone war erupted in North America that shows little indication of subsiding any time soon. The two groups involved in this war are not as easy to define as with the Marsh/Cope conflict, because there are many differing personalities involved on both sides. Nevertheless, I would identify the two combatant groups as the "independents" versus the "elites."

Most members of both groups are secular and hold strictly to the philosophical position of evolution and millions of years. However, their views on vertebrate fossil resources, morality, ethics, regulations, and legislation are in conflict, as are their views on what the proper expertise, experience, and education should be for those allowed to collect and study vertebrate fossils. The elites strongly believe that vertebrate fossils are a non-renewable resource that should be reserved for collection and study by only the few and kept ensconced in publically owned repositories. The independents believe that fossils are abundant and should be looked at more like minerals, gravel, or oil and should be harvested before weathering destroys them. As a creationist I think the secular "non-renewable" position supports the lack of evidence for evolution and uniformitarianism, since if those two presuppositionally-held positions were really true there should be a constantly-ongoing renewal of fossil resources in the rocks—and that is not reality. Most biblical creationists who study the sedimentary rock record see the vast majority of fossils as being a record of God's judgment brought through the one-time cataclysmic actions of the worldwide Flood.

So, the independent group of fossil hunters would like to be able to work within the American free enterprise system to discover, extract, study, describe, and publicize fossils. That is, as independent private companies they want to be free to make money in their endeavors. The Association of Applied Paleontological Sciences (AAPS) is a group organized to strengthen and publicize their position that the private (for-profit) collection of fossils can meet research and educational needs of the scientific community and the interested public just as well as can the elitist system.

On the other hand, the elite view is that the barter, sale, or purchase of scientifically significant vertebrate fossils should not be done unless accomplished within a public trust. This is the position taken by most government agencies that have control over public land, by many in academia, and by most secular natural history museums. Of course, the big disadvantage of this view is that the costs for vertebrate fossils accumulation is expected to be borne by the public (either directly or indirectly) using tax dollars, with the determination of what can be collected, studied, and displayed made only by those elites deemed qualified.

Personally, I think there is room for both fossil collecting methods, but every effort should be made to prevent those supporting the elite position from eradicating the free enterprise fossil collecting companies through onerous regulations and legislation.

It Is Very Hard Work!

No matter who does it, prospecting for, recovering, and collecting dinosaur fossils is very strenuous and difficult work, often accomplished in remote locations under extreme weather conditions. Some of the resources listed in the Bibliography go into great detail about dinosaur fossil collecting. See especially Larson and Donnan, 2004 and Sternberg,

Fig. VII–2: Digging Dinosaur Bones—Hot Work

Fig. VII–3: Digging Dinosaur Bones —Hard on the Back

1909 for good understanding of what fossil hunting is really like. Those who do this work on a regular basis must have a true love for the extraction process and the fossils they recover.

Preparation for Digging

The rules vary somewhat from place to place, but amateur collectors cannot recover or collect dinosaur bones (or any vertebrate fossils) from any public lands anywhere in North America. They could do so if they had a permit from the proper agency, but that is not likely to be forthcoming for amateurs. That limits collection to private property for most people other than those hunters who have the "acceptable" qualifications. Never trespass or collect fossils on private property without obtaining prior permission from the owner, and make sure you have done enough research to know who the owner really is. In this day and age it is advisable to get a written agreement with the owner, especially if you want to take fossils off the property. Expect to pay for the privilege if any of the fossils have much value.

I recommend too that anyone wanting to get serious about dinosaur bone collecting should read and follow the directives placed on the AAPS website before heading out. In fact, the best way to get started in the

Dinosaur Hunting

dino dig activities may be to sign up for an expert-led expedition before attempting anything on your own.

How Is It Done?

Generally speaking dinosaur bones are found in sedimentary layers in areas where the layers are not covered by much vegetation, so that weather can work to expose the "treasures" buried within. Oftentimes these locations are the badlands of North America like those shown in Figure VII–4.

Fig. VII–4: North American Badlands are likely places to discover dinosaur fossils.

Experienced dinosaur hunters usually look in or near locations where bones have been previously found or reported. The procedure often used is to slowly walk in gullies or other lower lying water-carrying locations with the head down and eyes focused on the ground, carefully watching for bone fragments. Once bone fragments are found the hunter then looks up higher in the sediments to see where the fragments might have originated. Perhaps a more significant find can be spotted and excavation can take place to save larger specimens before they disintegrate as a result of the actions of the environment.

Tools of the Trade

Novice dino hunters will want to have the following equipment and supplies in order to have much of a chance of completing successful fossil adventures:

- Planned route with applicable detailed maps. Possibly a GPS.
- Adequate food and water and a backpack.
- Proper permission to access the hunting area.
- Sun protection and insect repellant.
- Four-wheel drive vehicle with spare tires.
- Notebooks and pens/pencils for documenting the locations and extent of discovered fossils.
- Digital camera and extra batteries.
- Heavy work gloves and rugged shoes or boots.
- Picks, hammers, shovels, chisels etc., for excavating fossils.
- Whisk brooms, paint brushes, glues, and dental tools for exposing, stabilizing and cleaning fossils.
- A plan for and materials for protecting the fossils that may need to be removed such as tissue, aluminum foil, plaster of Paris, lock-top type plastic bags.

In a similar manner to many other types of endeavors, as experience is gained in dino hunting the list of required equipment will change and expand. After some experience, and in the case of fossils encased in hard matrix, power tools and heavy equipment may be needed to extricate the ancient treasures.

Section VIII

Dinosaur Paleontologists

THE MEN LISTED in this section all contributed greatly to the early discovery, description and promotion of North American dinosaurs. They laid the fossil material and theoretical foundations upon which the paleontologists that followed built their hypotheses and theories regarding dinosaurs.

Roy Chapman Andrews (1884–1960):

Worked for: The American Museum of Natural History, New York City

Noted for: Andrews was famous for his expeditions to the Gobi desert in the 1920s where new dinosaur genera were discovered and dinosaur eggs in a nest were first recovered.

Comments: Roy Chapman Andrews was not known for work on North American dinosaurs specifically, but as a swashbuckling adventurer complete with six-gun he was known world-wide as a tremendous promoter of dinosaurs. In the 1950s he wrote popular children's books on dinosaurs and "prehistoric" animals. Many have noticed the resemblances between Andrews and Indiana Jones, the fictional adventurer of the movies.

Barnum Brown (1873–1963):

Worked for: The American Museum of Natural History, New York City

Noted for: Brown was described as always being exquisitely dressed and perfectly proper whether he was in the office, the lab or the field. He discovered *Triceratops* in 1902 and *Tyrannosaurus* in 1908.

Comments: Barnum Brown was the museum's premier fossil hunter for over sixty years. He loved dinosaurs and was heavily involved in the museum's famous excavations at Como Bluff, Wyoming, in the late 1890s, as well as paleontological explorations of the Red Deer River in Alberta. He discovered in 1912 and named in 1914 *Anchiceratops* from that area of Canada and he was known to museum insiders as "Mr. Bones." Like Roy Chapman Andrews, Brown had a penchant for fame and publicity that resulted in him being described as "The world's most famous fossil hunter" in major metropolitan newspapers of the 1930s.

Edwin H. Colbert (1905–2001):

Worked for: The American Museum of Natural History, New York City and The Museum of Northern Arizona in Flagstaff

Noted for: Colbert was best known for his 1947 discovery of a large assemblage of *Coelophysis* dinosaur skeletons at Ghost Ranch in New Mexico. Even many secularists admit the probable cause for the demise of this batch of dinosaurs was a large flood.

Comments: Edwin H. Colbert was the dinosaur expert in residence at the American Museum from the 1930s through the 1950s and was instrumental in the opening of a new dinosaur hall there during that time. He wrote numerous popular and semi-technical publications up through the 1990s. In 1989 after 42 years of research he finally published his 160-page monograph on *Coelophysis*.

Edward Drinker Cope (1840–97):

Worked for: Mostly E.D. Cope worked only for himself. He managed to spend his inherited fortune on the search for and collection of dinosaur and other fossils.

Noted for: Cope is still famous for the "bone wars" between him and Othniel Charles Marsh, and a prolific amount of published material regarding evolution, vertebrate paleontology in general and dinosaurs specifically. He donated his brain to the Anthropometric Society for research concerning human intelligence.

Comments: Cope named a number of dinosaurs and, along with O.C. Marsh, managed to increase the number of published dinosaur genera twenty-fold during their decades of heated competition. In 1889 Cope named *Coelophysis*, a name still held today for this Lithe, Fast Runner. The dinosaur *Drinker* was named after Cope in 1990. Upon his death his extensive fossil collections were purchased by the American Museum of Natural History.

Earl Douglass (1862–1931):

Worked for: The Carnegie Museum of Natural History, Pittsburgh

Noted for: Douglass was best known for his 1908 discovery and subsequent excavation of the dinosaur-bone quarry now known as Dinosaur National Monument in northeastern Utah.

Comments: Douglass worked for thirteen years for the Carnegie Museum at the quarry extracting dinosaur bones and other fossils. A total of 350 tons of material was shipped from the site to the museum during that period. Douglass discovered the famous *Apatosaurus* that was fitted with the "wrong head" at this location. Twenty mountable dinosaur skeletons were gleaned from the rock sent to the museum including *Camarasaurus*, *Allosaurus*, and *Stegosaurus*. In 1915 Douglass wrote in his diary that he wished the quarry site would be set aside and preserved for public viewing. His dream was finally realized in 1958 when the National Park Service opened a permanent Quarry Visitor Center that allowed visitors to view hundreds of bones lying partially excavated in a long sandstone wall.

Charles W. Gilmore (1874–1945):

Worked for: The United States National Museum (Smithsonian Institution), Washington, D.C.

Noted for: Gilmore produced forty-three scientific papers and monographs on dinosaurs and undertook seven museum expeditions in search of dinosaur fossils.

Comments: After the Carnegie Museum ceased their work at Dinosaur National Monument because of a shortage of funds and storage space at the museum, Gilmore continued work there representing the Smithsonian. Among dinosaurs collected there by Gilmore was a *Diplodocus* skeleton that was put on display at the Smithsonian in 1931. During his career, Gilmore named numerous dinosaurs including *Alamosaurus*, *Brachyceratops*, and *Thescelosaurus*.

Lawrence Lambe (1863–1919):

Worked for: The Geological Survey of Canada

Noted for: Lambe brought the dinosaur-bearing sedimentary deposits along the Red Deer River in Alberta, Canada, to the attention of paleontologists elsewhere in the world

Comments: Lambe was the first Canadian to attempt scientific descriptions of Red Deer River fossils. He worked with Henry Fairfield Osborn to describe the bones he and his survey crew had collected on several expeditions around the turn of the twentieth century. He named several genera of Horn-Faced dinosaurs including *Centrosaurus* in 1904, *Styracosaurus* in 1913, and *Chasmosaurus* in 1914. The Duck-Bill *Lambeosaurus* was named in honor of Lambe in 1923 by W. A. Parks.

Joseph Leidy (1823–91):

Worked for: Academy of Natural Sciences, Philadelphia

Noted for: Leidy was the supreme American consultant in numerous scientific fields for five decades during the 19th century, and is credited for being the founder of American paleontology.

Comments: Until E.D. Cope and O.C. Marsh began paying for fossils in the early 1870s, Leidy was the primary person American naturalists looked to for scientific descriptions of fossils of all kinds. He was a hardworking generalist who preferred to describe rather than theorize about his science. Leidy concentrated on paleontology from 1856 to 1872, and between the late 1850s and 1870 he examined almost every fossil found in the western United States. Altogether he published over 200 papers on paleontology totaling over 1,800 pages.

Richard S. Lull (1867–1957):

Worked for: Peabody Museum at Yale University, New Haven

Noted for: His study of dinosaur bones and tracks of the Connecticut Valley

Comments: There are thousands of dinosaur footprints that have been found in the Connecticut Valley, but very few dinosaur skeletal remains have been discovered. The fossil fragments that have been found were given genus names of *Anchisaurus* and *Ammosaurus* by O.C. Marsh in the late 1800s. But Marsh, Lull and those paleontologists that followed them relied heavily on the plentiful European dinosaur *Plateosaurus* for their visualization of the east coast dinosaurs. That is because the few American fossils fragments recovered are similar to those of *Plateosaurus*. Professor Lull also wrote technical articles about dinosaurs found in other locations in North America such as *Triceratops*, *Stegosaurus*, *Barosaurus*, and *Camarasaurus*.

Othniel Charles Marsh (1831–99):

Worked for: Peabody Museum at Yale University, New Haven

Noted for: Marsh is still famous for the "Bone wars" with E.D. Cope and for his contributions to evolutionary paleontology. He was also known for his astonishing collection of fossils that he collected over a thirty year period and that are now on display or housed at Peabody Museum and the Smithsonian.

Comments: O.C. Marsh was a nephew of millionaire George Peabody and received a large amount of monetary assistance from Peabody. This assistance allowed for, among other things, Marsh's ability to pay men to find and recover fossils for him to prepare and describe. Marsh published skeletal reconstructions in the 1880s and 1890s of *Brontosaurus* (now *Apatosaurus*), *Stegosaurus, Triceratops, Anchisaurus, Camptosaurus,* and several others. He worked tirelessly to promote Darwinian evolution, although he never was able to formulate himself any evolutionary generalization for what he found in his vast collection of fossils.

Henry Fairfield Osborn (1857–1935):

Worked for: The American Museum of Natural History, New York City

Noted for: Osborn was the first curator of vertebrate paleontology and longtime president at the American Museum. He exhibited far-reaching influence on paleontology between 1890 and 1930 through his promotional and organizational abilities at the museum.

Comments: Osborn was born into a very wealthy family and developed a confidence in his abilities that allowed him to propose numerous evolutionary theories while watching over an ambitious schedule of field expeditions. He directed fossil hunts that resulted in dinosaur displays that helped bring the American Museum to worldwide fame. He used many of his personal funds to assist in the expansion of many displays there. Osborn was a devoted follower of E.D. Cope.

Charles H. Sternberg (1850–1943):

Worked for: Sternberg spent many years working directly for E.D. Cope. He also collected for the Geological Survey of Canada and others.

Noted for: Sternberg had a life-long career as a fossil hunter. He helped pioneer the Canadian boom in dinosaur discovery and excavation from 1912–1917.

Comments: Along with his three sons (George, Charles M., and Levi), Charles H. excavated fossils throughout North America. Their fossils, including many dinosaur fossils, were sent to natural history museums all over the world. The Sternbergs and Barnum Brown developed a short-lived competition for dinosaur bones in Alberta similar to that of Cope and Marsh. Concentrated in the area now encompassed by Dinosaur Provincial Park, the "Great Canadian Dinosaur Rush" of 1914 pitted bone collectors from the American Museum of Natural History (led by Brown) against those from the Geological Survey of Canada (led by George Sternberg).

Section IX

Directory of Created Kinds in Alphabetical Order:

1. Armor-Backed Dinosaurs — *Edmontonia*
2. Bipedal Browser Dinosaurs — *Camptosaurus*
3. Club-Tailed Dinosaurs — *Ankylosaurus*
4. Duck-Billed Dinosaurs — *Edmontosaurus*
5. Horn-Faced Dinosaurs — *Triceratops*
6. Horn-Nosed Bipedal Dinosaurs — *Ceratosaurus*
7. Killer-Clawed Dinosaurs — *Deinonychus*
8. Lithe, Fast Running Dinosaurs — *Coelophysis*
9. Long-Necked Big-Clawed Dinosaurs — *Anchisaurus*
10. Long-Necked Boxy-Headed Dinosaurs — *Brachiosaurus*
11. Long-Necked Slender-Headed Dinosaurs — *Diplodocus*
12. Ostrich-Like Dinosaurs — *Ornithomimus*
13. Plate-Backed Dinosaurs — *Stegosaurus*
14. Thick-Headed Dinosaurs — *Pachycephalosaurus*
15. Tyrant Bipedal Dinosaurs — *Tyrannosaurus*

Section IX

Dinosaur Data by Created Kind

Keys to Understanding the Dinosaur Classifications

FOR EACH CREATED kind classification in the guidebook there are fifteen information sections. Explanations for the information in these sections are as follows:

1. **Created Kind**: This is the descriptive name applied for the kind. Some of these names are so descriptive that they were probably initially coined by others; some other names may have been originated by the author and are thus unique to this guidebook.
2. **Representative Genus Name and Meaning**: In most cases the representative genus name chosen is the one for which most North American fossil material has been discovered, prepared, described and displayed. A pronunciation help for the name is provided along with the meaning of the name and the source language(s) for the scientific name. The representative genus name should usually be familiar to the general public, but this was not a prerequisite for the naming process. It is logical to expect that the dinosaurs for which the most fossil material has been recovered would also be most famous.
3. **Day Created**: The Bible explains in Genesis that God made animals on day six of the creation. Dinosaurs are animals, so they were all created on day six. The initially-created dinosaurs were endowed with DNA that allowed for the genetic variation seen in the animals recovered as fossils from the rock record due to the Flood. Dinosaurs came off the Ark after the Flood as

well, but they seem to have mostly, if not all, gone extinct. It is hypothesized that we have not found remnants of the post-Flood dinosaurs in the North American rock record.

4. **Hip Design**: Dinosaurs have been discovered as fossils with two different types of hip designs. All dinosaurs can be assigned to one or the other of these hip-design categories. They are either Bird-Hipped (Ornithischian) or Lizard-Hipped (Saurischian). See also Section II.

5. **Teeth Design**: A description of the teeth of each kind is provided in this section. The Bible says that in the beginning all animals ate plants, and so in the beginning dinosaurs all ate plants. After the Fall indications from fossils are that some animals began to eat other animals even before the Flood.

6. **Skull Design**: Dinosaurs had a number of different skull designs just as do the existing animals of today. It can be assumed that there was variation within each kind for dinosaurs so far as skull design and size, just as there are with extant animals we see today.

7. **Feet Design**: This is basically a description of the claws and the numbers of toes and fingers on each foot or hand.

8. **Stance Design**: Examination of the fossil skeletons of dinosaurs has led science to determine that they were either bipedal (walking on two feet) or quadrupedal (walking on four feet). In some kinds it is speculated that they may have used both means at times, like a bear for example.

9. **Size**: Estimated maximum overall length is given in feet. The lengths have been determined from the actual or theoretical reconstruction of skeletal parts into a complete animal. Weight estimates are not given in this guidebook due to the lack of information from the fossils on the animal body shapes, internal organs, tissue types, and other aspects of the dinosaurs when they were alive. There are guidebooks that do provide some weight estimates for some genera, but they have been developed using highly conjectural assumptions. In some genera non-adult specimens are thought to have been discovered that are smaller than the largest or average-sized specimens. And as with extant animals it would be a logical assumption that there would have been size variation due to differences between males and females (dimorphism).

10. **States (Provinces) Where Fossils Have Been Found**: Since this is a guidebook of North American dinosaurs the American states and Canadian provinces are listed where fossils have been

Dinosaur Data by Created Kind

discovered. There has been some paleontological work done in Mexico, but the dinosaur fossils prepared and described from there are minimal at the time of publication of this guidebook.

11. **When First Discovered, Where, and by Whom:** Some basic history for the particular dinosaur kind is given in this section of the guidebook.
12. **Extent of Fossils Found:** The reader may be surprised to discover how little fossil material has been discovered and described for several of the dinosaur kinds.
13. **Synonyms (Other likely North American genera for the same dinosaur kind):** The created kinds for dinosaurs probably settle in at approximately the Linnaean "Family" classification. With the move by secular paleontology to cladistics, any attempt to use cladograms to develop created kinds is futile. The genera included in each created kind in the guidebook have been established due to analyses of animal form, function, and structure. Name pronunciation of the scientific names and sources for the names are also provided for each genus in this section.
14. **Other Interesting Facts:** These are the facts that were of most interest to the author and it is hoped that many readers will concur.
15. **Artistic Reconstruction:** As mentioned earlier, scientists do not know exactly what the living dinosaurs looked like. Perhaps someday dinosaurs will be found alive in some remote location of the earth. Nevertheless, until then artists will continue to use their imaginations to speculate on what dinosaurs may have looked like in the flesh.
16. **Photographs:** All of the photographs in each grouping are of appropriate fossil and cast material, mostly from secular natural history museums. An attempt has been made to fairly represent the kinds and the quantities of displays from these museums. Keep in mind that not all museums identify which of their displays consist of replica casts, and as mentioned previously, even the more complete skeletal reconstructions may be made of a considerable quantity of material that is not original. The photos in this section are important because they illustrate the material that the secular paleontological community has chosen in order to sell to the public the philosophy of evolution and millions of years. It is also the material that creationists heavily rely on to educate those willing to consider the truth of the biblical perspective on dinosaurs.

Dinosaur Classification

CREATED KIND #1 – ARMOR-BACKED DINOSAURS

1. **Created Kind:** Armor-Backed Dinosaurs
2. **Representative Genus Name and Meaning:** *Edmontonia* (ED-mon-TOH-nee-ah). From Edmonton (English).
3. **Day Created:** Six.
4. **Hip Design:** Bird-Hipped (Ornithischian).
5. **Teeth Design:** Cropping beak with rows of small, serrated teeth.
6. **Skull Design:** Broad with top armor plates.
7. **Feet Design(s):** Three or four toes.
8. **Stance Design:** Quadrupedal.
9. **Size:** Up to 25 feet long.
10. **States (Provinces) Where Fossils have been Found:** AK, KS, MT, SD, UT, WY, TX (AB).
11. **When First Discovered, Where, and by Whom:** *Panoplosaurus* was first discovered in 1919 in Alberta and named by L. Lambe. *Edmontonia* was also found in Alberta and was named by Charles Sternberg in 1928.
12. **Extent of Fossils Found:** There is no record of a complete armor-backed skeleton having been found at a location, but several partial skeletons and over a dozen skulls are reported.
13. **Synonyms (Other likely North American genera for the same dinosaur kind):**

    Animantarx (an-ee-MAN-tarks). Animated fortress (Latin).

    Gastonia (GAS-tone-ee-uh). Gaston's animal (Greek).

    Niobrarasaurus (NYE-oh-bra-rah-SAW-rus). Niobrara chalk lizard (English/Greek).

    Panoplosaurus (PAN-op-loh-SAW-rus). All-armored lizard (Greek).

    Pawpawsaurus (paw-paw-SAW-rus). Paw Paw formation lizard (English/Greek).

    Sauropelta (SAWR-oh-PEL-tuh). Lizard shield (Greek).

    Silvisaurus (SIL-vuh-SAW-rus). Forest lizard (Latin/Greek).

14. **Other Interesting Facts:** The armor-backed dinosaurs had armor that consisted of spines and spikes. They were similar to the tail-clubbed dinosaurs except no individuals have yet been discovered with clubs on their tails. Some paleontologists believe studies of the skulls indicated *Edmontonia* could make honking noises.

Dinosaur Data by Created Kind 83

15. **Artistic Reconstruction:**

Fig. IX–1: Kind #1: Armor-Backed Dinosaur
Artistic Reconstruction by M. Pike

16. **Photographs:**

Fig. IX–1.1: *Animantarx* Skeletal Reconstruction (Price, Utah Museum)

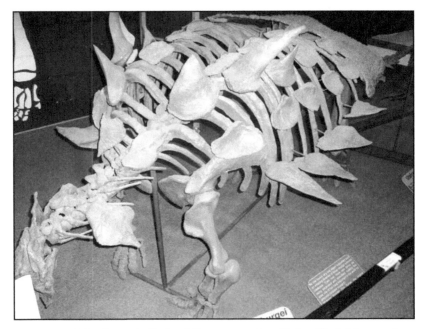

Fig. IX–1.2: *Gastonia* Skeletal Reconstruction (Price, Utah Museum)

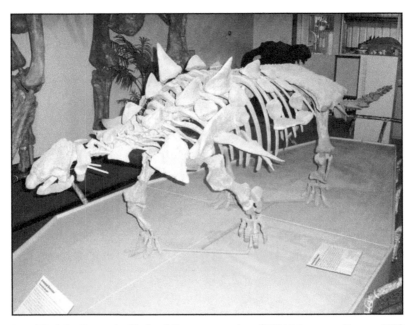

Fig. IX–1.3: *Gastonia* Skeletal Reconstruction (BYU Museum, Provo, UT)

Dinosaur Data by Created Kind 85

Fig. IX–1.4: Armor-Backed Skeletal Reconstruction from the Morrison Formation in Colorado (WDC)

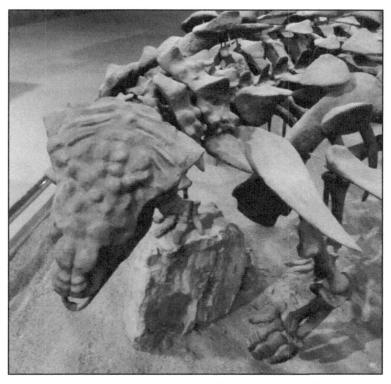

Fig.IX–1.5: Armor-Backed Skeletal Reconstruction from the Morrison Formation in Colorado (WDC)

Fig. IX–1.6: Armor-Backed Skull Reconstruction from the Morrison Formation in Colorado (WDC)

Fig. IX–1.7: Armor-Backed Skeletal Reconstruction from the Morrison Formation in Colorado (WDC)

Fig. IX–1.8: *Gastonia* Skeletal Reconstruction at Wyoming Dinosaur Center

Fig. IX–1.9: *Gastonia* Skeletal Reconstruction at Wyoming Dinosaur Center

Dinosaur Classification
CREATED KIND #2 – BIPEDAL BROWSER DINOSAURS

1. **Created Kind:** Bipedal Browser Dinosaurs
2. **Representative Genus Name and Meaning:** *Camptosaurus* (KAMP-toh-SAW-rus). Flexible lizard (Greek).
3. **Day Created:** Six.
4. **Hip Design:** Bird-Hipped (Ornithischian).
5. **Teeth Design:** Sharp, horny beak; hundreds of small chisel-like teeth.
6. **Skull Design:** Long and fairly narrow.
7. **Feet Design(s):** Four-toed hind feet, five-fingered stout hands and arms.
8. **Stance Design:** Bipedal, but probably browsed on all fours.
9. **Size:** Up to 23 feet long.
10. **States (Provinces) Where Fossils Have Been Found:** CO, MT, OK, TX, UT, WY
11. **When First Discovered, Where, and by Whom:** First described by O.C. Marsh in 1885 and first discovered in the late 1870s in Wyoming.
12. **Extent of Fossils Found:** Over 30 partial articulated skeletons and over 25 disarticulated skull elements mostly from *Camptosaurus* and *Tenontosaurus*.
13. **Synonyms (Other likely North American genera for the same dinosaur kind):** *Drinker* (DREEN-kur). Edward Drinker Cope's one (English).

    Orodromeus (oro-DROM-ee-us). Mountain runner (Greek).

    Othnielia (oth-NEEL-yuh). Othniel Charles Marsh's one (English/Greek).

    Parkosaurus (PARK-oh-SAW-rus). William Park's lizard (English/Greek).

    Tenontosaurus (TEN-on-toh-SAW-rus). Tendon lizard (Greek).

    Thescelosaurus (THESS-el-oh-SAW-rus). Marvelous lizard (Greek).
14. **Other Interesting Facts:** These bipedal browser dinosaurs are similar to the *Iguanodon* that is well known for being one of the first dinosaurs scientifically described (in Belgium).

15. Artistic Reconstruction:

Fig. IX–2: Kind #2: Bipedal Browser Dinosaur, Artistic Reconstruction by M. Pike

16. Photographs:

Fig. IX–2.1: *Camptosaurus* Skeletal Reconstruction (Price, Utah Museum)

Fig. IX–2.2: *Camptosaurus* Skeletal Reconstruction
(BYU Museum Provo, Utah)

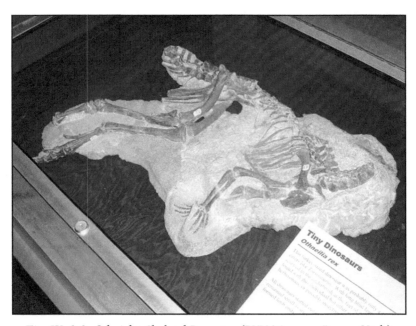

Fig. IX–2.3: *Othnielia* Skeletal Remains (BYU Museum Provo, Utah)

Dinosaur Data by Created Kind

Fig. IX–2.4: *Othnielia* Skeletal Reconstruction at Wyoming Dinosaur Center

Fig. IX–2.5: *Othnielia* Skeletal Reconstructions at Wyoming Dinosaur Center, Thermopolis, WY.

Fig. IX–2.6: *Tenontosaurus* Skull at Ft. Worth Museum of Science and Industry

Fig. IX–2.7: *Tenontosaurus* Skeletal Reconstruction at Ft. Worth Museum

Fig. IX–2.8: *Tenontosaurus* Skeletal Reconstruction at Ft. Worth Museum

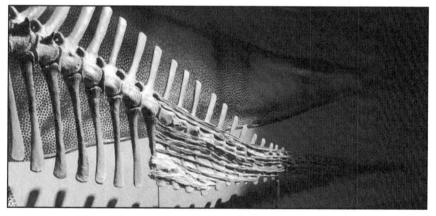

Fig. IX–2.9: *Tenontosaurus* Tail with Ossified Tendons at Ft. Worth Museum

Fig. IX–2.10: *Tenontosaurus* Fossilized Foot at Ft. Worth Museum

Fig. IX–2.11: *Tenontosaurus* Fossilized Foot at Ft. Worth Museum

Dinosaur Classification

CREATED KIND #3 – CLUB-TAILED DINOSAURS

1. **Created Kind:** Club-Tailed Dinosaurs
2. **Representative Genus Name and Meaning:** *Ankylosaurus* (an-KEE-loh-SAW-rus). Curved lizard (Greek).
3. **Day Created:** Six.
4. **Hip Design:** Bird-Hipped (Ornithischian).
5. **Teeth Design:** Toothless beak and small teeth.
6. **Skull Design:** Compact head, rounded and armored.
7. **Feet Design(s):** Three or four toes on each foot.
8. **Stance Design:** Quadrupedal.
9. **Size:** Up to 35 feet long.
10. **States (Provinces) Where Fossils Have Been Found:** MT, TX, WY, (AB)
11. **When First Discovered, Where, and by Whom:** Genus *Ankylosaurus* was found by Barnum Brown in Montana in 1908. Lawrence Lambe named *Euoplocephalus* from remains found in Alberta in 1910.
12. **Extent of Fossils Found:** About 20 complete or partial skulls, a few partial skeletons, and one nearly complete skeleton have been found.
13. **Synonyms (Other likely North American genera for the same dinosaur kind):** *Euoplocephalus* (yoo-oh-ploh-SEF-uh-lus). Well-armed head (Greek).
14. **Other Interesting Facts:** The tail clubs and armor on these animals are thought by paleontologists to have provided serious defense against attackers. Researchers have discovered that the skulls of these animals had a complex system of passages in their nasal passages, but there is no consensus on the purpose for them. It is believed by some that the club-tailed dinosaurs even had bony eyelids.

15. Artistic Reconstruction:

Fig. IX–3: Kind #3: Club-Tailed Dinosaur, Artistic Reconstruction by M. Pike

16. Photographs:

Fig. IX–3.1: *Euoplocephalus* Skeletal Reconstruction (Royal Tyrell Museum, Alberta)

Dinosaur Data by Created Kind 97

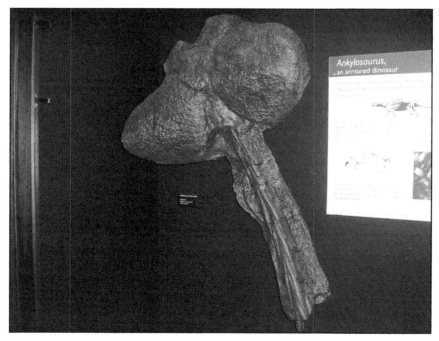

Fig. IX–3.2: *Ankylosaurus* Fossil Club (Museum of the Rockies, Montana)

Dinosaur Classification

CREATED KIND #4 – DUCK-BILLED DINOSAURS

1. **Created Kind:** Duck-Billed Dinosaurs
2. **Representative Genus Name and Meaning:** *Edmontosaurus* (ed-MONT-oh-SAW-rus). Edmonton lizard (English/Greek).
3. **Day Created:** Six.
4. **Hip Design:** Bird-Hipped (Ornithischian).
5. **Teeth Design:** Horny, toothless beak, tightly-packed rows of tiny leaf-shaped teeth.
6. **Skull Design:** Elongated with crests.
7. **Feet Design(s):** Three toes on hind feet, 3 or 4 toes on front feet.
8. **Stance Design:** Quadrupedal, but able to stand on hind legs.
9. **Size:** Up to 45 feet long.
10. **States (Provinces) Where Fossils Have Been Found:** CO, MT, NJ, NM, ND, WY, (AB, SK)
11. **When First Discovered, Where, and by Whom:** The first North American duck-billed dinosaur (*Hadrosaurus*) was discovered in New Jersey and described by Joseph Leidy in 1858. In the late 1800s E.D. Cope and O.C. Marsh described the fossils from a number of duck-billed dinosaurs from the American West and Alberta, Canada.
12. **Extent of Fossils Found:** The fossil material found for the duck-bills is relatively extensive. At least 75 complete or nearly complete skulls have been found along with about the same number of disarticulated or partial skulls. Probably fewer than six complete skeletons are reported found along with about 20 partial skeletons.
13. **Synonyms (Other likely North American genera for the same dinosaur kind):** *Anatotitan* (ANN-at-oh-TYTE-an). Giant duck (Latin/Greek).

    Brachylophosaurus (BRAK-ee-LOF-oh-SAW-rus). Short-crest lizard (Greek).

    Corythosaurus (koh-RITH-oh-SAW-rus). Helmeted lizard (Greek).

    Gryposaurus (GRI-poh-SAW-rus). Hook-nosed lizard (Greek).

    Hadrosaurus (HAD-roh-SAW-rus). Strong lizard (Greek)

    Hypacrosaurus (HIP-AK-crow-SAW-rus). Below highest lizard (Greek).

Kritosaurus (CRY-toh-SAW-rus). Separated lizard (Greek).

Lambeosaurus (LAM-bee-oh-SAW-rus). Lambe's lizard (English/Greek).

Maiasaura (MY-uh-saw-rah). Good mother lizard (Greek).

Parasaurolophus (PAR-uh-saw-roh-LOW-fus). Beside *Saurolophus* (Greek).

Prosaurolophus (pro-SAW-roh-LOW-fus). Before *Saurolophus* (Greek).

Saurolophus (SAW-roh-LOW-fus). Lizard crest (Greek).

14. **Other Interesting Facts:** At one time paleontologists thought the duck-billed dinosaurs spent most of their time in the water to evade predators and to eat water plants. The current popular consensus is that they mostly lived on land and were not good swimmers.
15. **Artistic Reconstruction:**

Fig. IX–4: Kind #4: Duck-Billed Dinosaur, Artistic Reconstruction by M. Pike

16. Photographs:

Fig. IX–4.1: *Lambeosaurus* Skeletal Reconstruction
(Carnegie Museum, Pittsburgh)

Fig. IX–4.2: *Prosaurolophus* Skeletal Reconstruction (Price, Utah Museum)

Fig. IX–4.3: *Prosaurolophus* Rear Foot (Price, Utah Museum)

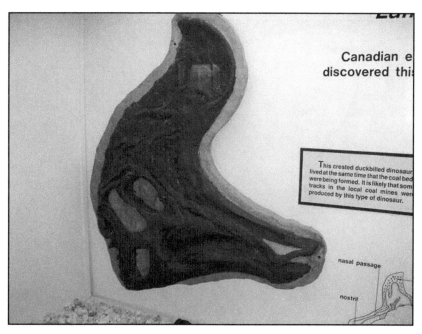

Fig. IX–4.4: *Lambeosaurus* Skull (Price, Utah Museum)

Fig. IX–4.5: Duck-Billed Dinosaur Tibia Bone (Museum of the Rockies, Montana)

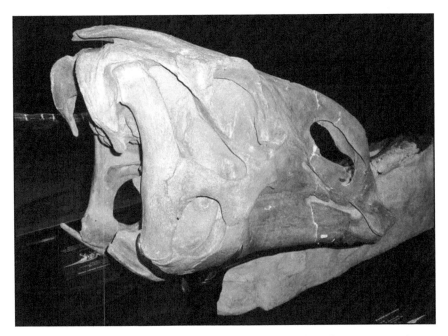

Fig. IX–4.6: *Brachylophosaurus* Skull (Museum of Rockies, Montana)

Dinosaur Data by Created Kind 103

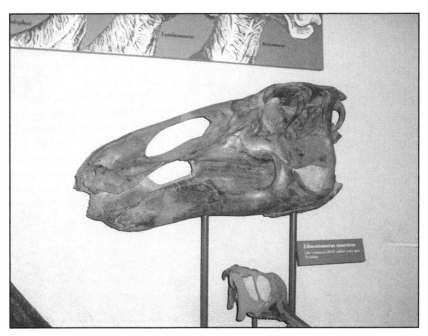

Fig. IX–4.7: *Edmontosaurus* Skull (Smithsonian Museum, Washington, D.C.)

Fig. IX–4.8: *Lambeosaurus* Skeletal Reconstruction (Tyrrell Museum, Alberta)

Fig. IX–4.9: *Lambeosaurus* Skeletal Reconstruction (Tyrrell Museum, Alberta)

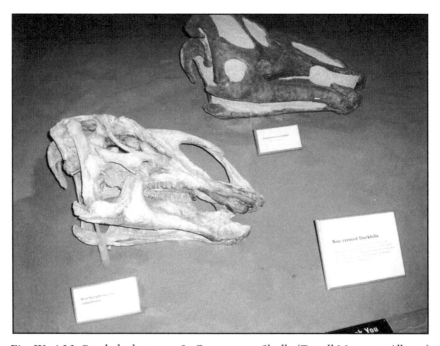

Fig. IX–4.10: *Brachylophosaurus* & *Gryposaurus* Skulls (Tyrrell Museum, Alberta)

Dinosaur Data by Created Kind 105

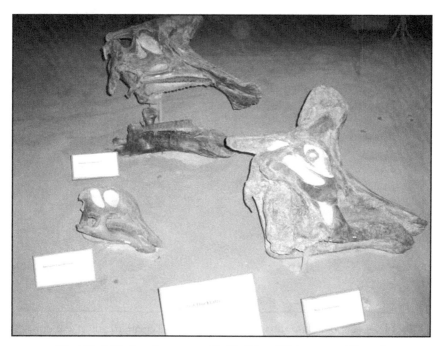

Fig. IX–4.11: *Lambeosaurus* Skulls (Tyrrell Museum, Alberta)

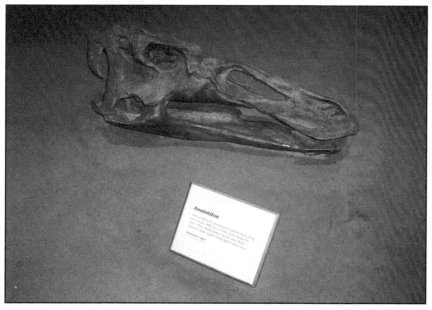

Fig. IX–4.12: *Anatotitan* Skull (Tyrrell Museum, Alberta)

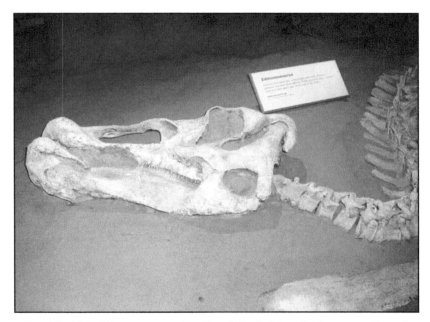

Fig. IX–4.13: *Edmontosaurus* Skull and Neck (Tyrrell Museum, Alberta)

Fig. IX–4.14: Duck-Billed Dinosaur Mummy, Showing Skin (San Diego Museum)

Fig. IX–4.15: Close-up of Duck-Billed Dinosaur Fossilized Skin (San Diego Museum)

Fig. IX–4.16: *Edmontosaurus* Replica Skull from Hell Creek Formation, South Dakota (Black Hills Institute)

Fig. IX–4.17: *Maiasaura* Replica Skull from Two Medicine Formation, Montana (Black Hills Institute)

Fig. IX–4.18: *Edmontosaurus* Skull on Skeletal Reconstruction from South Dakota (Black Hills Institute)

Fig. IX–4.19: *Edmontosaurus* Skeletal Composite Mount from Hell Creek Formation, SD (SDMG)

Fig. IX–4.20: *Edmontosaurus* Skull in Composite Mount from Hell Creek Formation, SD (SDMG)

Fig. IX–4.22: *Edmontosaurus* Hip/Femurs in Composite Mount Hips/Femurs from Hell Creek Fm. (SDMG)

Fig. IX–4.21: Edmontosaurus Foot in Composite Mount from Hell Creek Formation, SD (SDMG)

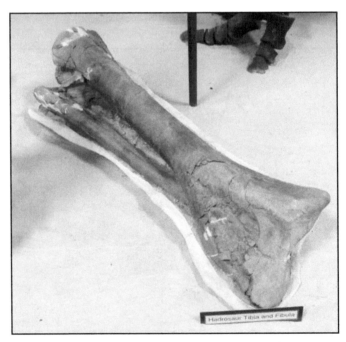

Fig. IX–4.23: *Hadrosaur* Tibia and Fibula (South Dakota Museum of Geology)

Fig. IX–4.24: *Maiasaura* Skeletal Reconstruction at Wyoming Dinosaur Center, Thermopolis, WY

Fig. IX–4.25: *Maiasaura* Skeletal Reconstruction at Wyoming Dinosaur Center

Fig. IX–4.26: *Maiasaura* Replica Skull Reconstruction at Wyoming Dinosaur Center in Thermopolis, WY

Dinosaur Classification

CREATED KIND #5 – HORN-FACED DINOSAURS

1. **Created Kind:** Horn-Faced Dinosaurs
2. **Representative Genus Name and Meaning:** *Triceratops* (try-SAIR-uh-TOPS). Three-horned face (Latin/Greek).
3. **Day Created:** Six.
4. **Hip Design:** Bird-Hipped (Ornithischian)
5. **Teeth Design:** Parrot-like beak, chisel-like teeth in batteries with many teeth ready to erupt from the jaw.
6. **Skull Design:** Heavy, armored head with frill.
7. **Feet Design(s):** Four splayed toes on each hooved foot.
8. **Stance Design:** Quadrupedal.
9. **Size:** Up to 30 feet long.
10. **States (Provinces) Where Fossils Have Been Found:** AK, CO, MT, NM, SD, TX, UT, WY, (AB, SK)
11. **When First Discovered, Where, and by Whom:** *Monoclonius* by E.D. Cope in 1876 from fossils found in Montana; *Triceratops* by O.C. Marsh in 1889 in Wyoming.
12. **Extent of Fossils Found:** Whereas skulls are hard to find for some kinds of dinosaurs, they are a common type of find for the horn-faced kind. Over 150 complete or partial skulls are reported found in North America. Several skeletons of the horn-faced kind have also been discovered.
13. **Synonyms (Other likely North American Genera for the Same Dinosaur Kind):** *Achelousaurus* (AK-uh-loo-SAW-rus). Lizard of Acheloo (Greek).
    Anchiceratops (an-kee-SAIR-uh-TOPS). Near-horned face (Greek).
    Arrhinoceratops (are-RYE-no-SAIR-uh-TOPS). No nose-horn face (Greek).
    Avaceratops (AVE-uh-SAIR-uh-TOPS). Ava's horned-face (English/Greek).
    Brachyceratops (BRAK-ee-SAIR-uh-TOPS). Short-horned face (Greek).
    Centrosaurus (SENT-roh-SAW-rus). Sharp-point lizard (Greek).
    Chasmosaurus (KAZ-moh-SAW-rus). Chasm lizard (Greek).
    Diceratops (dy-SAIR-uh-TOPS). Two-horned face (Greek).
    Einiosaurus (EYE-knee-oh-SAW-rus). Bison lizard (Blackfoot/Greek).
    Leptoceratops (lep-toe-SAIR-uh-TOPS). Slender-horned face (Greek).
    Monoclonius (MON-oh-KLON-ee-us). Single twig (Latin/Greek).

Montanoceratops (MON-tan-oh-SAIR-uh-TOPS). Montana horned-face (English/Greek).

Pachyrhinosaurus (pack-ee-RYE-no-SAW-rus). Thick-nosed lizard (Greek).

Pentaceratops (pent-uh-SAIR-uh-TOPS). Five-horned face (Greek).

Styracosaurus (sty-RACK-oh-SAW-rus). Spike lizard (Greek).

Torosaurus (TOR-oh-SAW-rus). Pierced lizard (Greek).

Zuniceratops (ZOON-ee-SAIR-uh-TOPS). Zuni tribe horned face (Greek).

14. **Other Interesting Facts:** The horn-face fossils vary in the shape and sizes of the horns, skull and frills, while the skeletons are very similar. Paleontologists do not agree on the orientation of the legs with respect to the body. Some think the front legs were splayed outward and the rear legs were more directly under the animal like in a Rhinoceros. Others believe all four legs were directly under the animal.

    The head is always the most noticeable feature on these animals, with the length of one *Triceratops* head measured to be nearly seven feet long. Some skin impressions of the horned faces have been found that indicate they had scales on their skin. The heads were connected to the skeleton with a ball-and-socket joint in the neck (at the balance point) that allowed the massive head to swivel.

15. **Artistic Reconstruction:**

Fig. IX–5: Kind #5: Horn-Faced Dinosaur, Artistic Reconstruction by M. Pike

16. Photographs:

Fig. IX–5.1: *Triceratops* Skull (Carnegie Museum, Pittsburgh)

Fig. IX–5.2: *Chasmosaurus* Skeletal Reconstruction (Price, Utah Museum)

Dinosaur Data by Created Kind 115

Fig. IX–5.3: *Triceratops* Skeletal Reconstruction (Tyrrell Museum, Alberta)

Fig. IX–5.4: *Chasmosaurus* Skeletal Reconstruction (Tyrrell Museum, Alberta)

Fig.IX–5.5: *Pachyrhinosaurus* Skeletal Reconstruction (Tyrrell Museum, Alberta)

Fig. IX–5.6: *Triceratops* Skeletal Reconstruction (Tyrrell Museum, Alberta)

Dinosaur Data by Created Kind 117

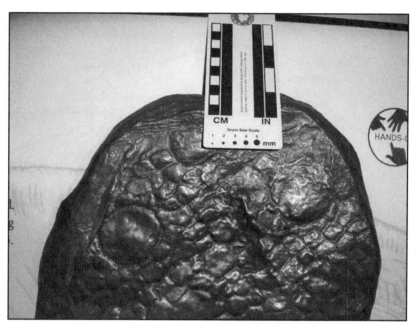

Fig. IX–5.7: *Centrosaurus* Skin Impression (Tyrrell Museum, Alberta)

Fig. IX–5.8: Parrot-like beak from Horn-Faced Dinosaur
(Tyrrell Museum, Alberta)

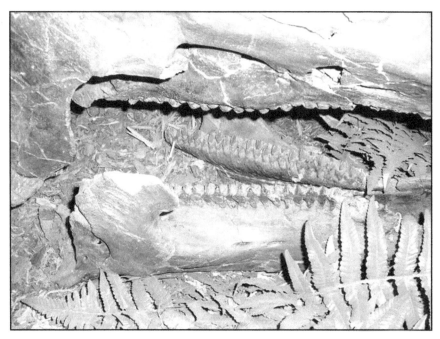

Fig. IX–5.9: Close-up of Teeth in Horn-Faced Dinosaur Jaws (Glendive, MT)

Fig. IX–5.10: *Monoclonius* Skull (BYU Museum Provo, Utah)

Fig.IX–5.11: *Monoclonius* Skull – See Ball for Socket
(BYU Museum Provo, Utah)

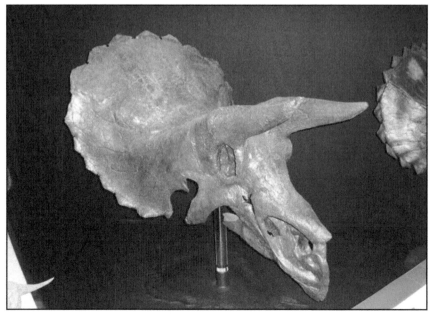

Fig. IX–5.12: *Triceratops* Skull (Museum of Rockies, Montana)

Fig. IX–5.13: *Triceratops* Skull (Museum of Rockies, Montana)

Fig. IX–5.14: *Triceratops* Skull (Museum of Rockies, Montana)

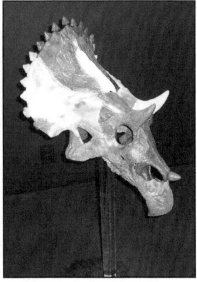

Fig. IX–5.15: *Triceratops* Skull (Museum of Rockies, Montana)

Dinosaur Data by Created Kind

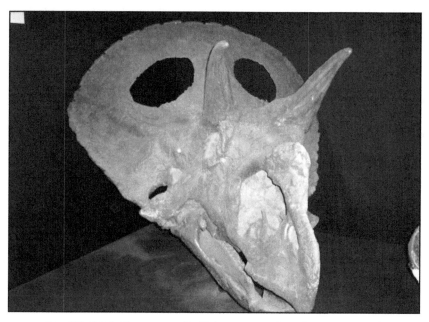

Fig. IX–5.16: *Torosaurus* Skull (Museum of Rockies, Montana)

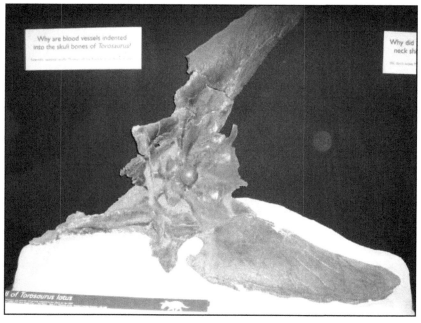

Fig.IX–5.17: Rear of *Torosaurus* Skull with Ball for Socket
(Museum of Rockies, Montana)

Fig. IX–5.18: *Triceratops* Skull – Baby? (Museum of Rockies, Montana)

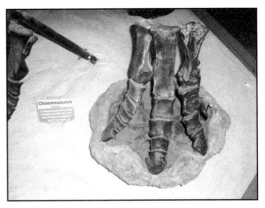

Fig. IX–5.19: *Chasmosaurus* Hind Foot Reconstruction (Price, Utah Museum)

Fig. IX–5.20: *Chasmosaurus* Front Foot Reconstruction (Price, Utah Museum)

Dinosaur Data by Created Kind 123

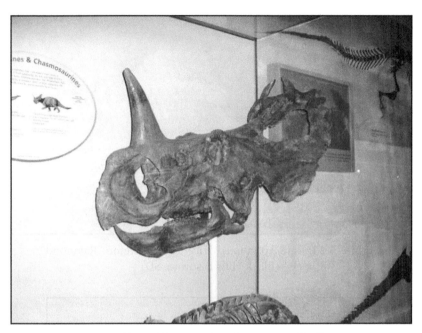

Fig. IX–5.21: Horn-Faced Dinosaur Skull
(Smithsonian Museum, Washington, D.C.)

Fig. IX–5.22: *Triceratops* Reconstructed Skull from Bone
BHI-4772 Hell Creek Fm., SD (Black Hills Institute)

Fig. IX–5.23: Rare Articulated *Triceratops* Skeleton "Raymond" (Black Hills Institute, SD)

Fig. IX–5.24: *Triceratops* Skull from Hell Creek Formation, SD (South Dakota Museum of Geology)

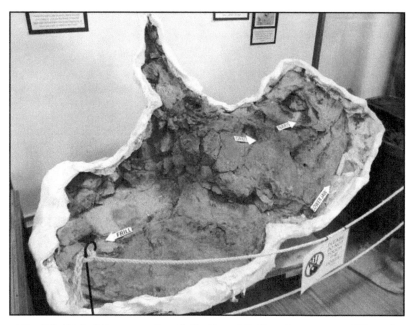

Fig. IX–5.25: *Triceratops* Fossil Skull and Frill in Preparation (SDMG)

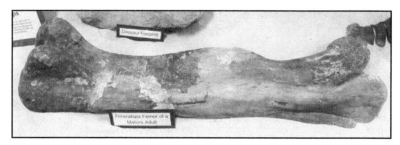

Fig. IX–5.26: *Triceratops* Femur (South Dakota Museum of Geology)

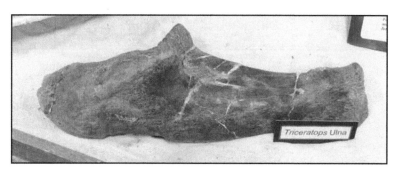

Fig. IX–5.27: *Triceratops* Ulna (South Dakota Museum of Geology)

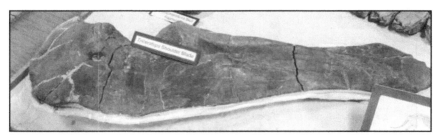

Fig. IX–5.28: *Triceratops* Shoulder Blade (Scapula)
(South Dakota Museum of Geology)

Fig. IX–5.29: *Triceratops* Skull Reconstruction at Wyoming
Dinosaur Center, Thermopolis

Dinosaur Data by Created Kind 127

Fig. IX–5.30: *Triceratops* Skeletal Reconstruction at
Wyoming Dinosaur Center

Fig. IX–5.31: *Zuniceratops* Skeletal Reconstruction from New Mexico
(Wyoming Dinosaur Center)

Fig. IX–5.32: *Zuniceratops* Skull Reconstruction from New Mexico (Wyoming Dinosaur Center)

Fig. IX–5.33: *Zuniceratops* Skeletal Reconstruction from New Mexico (Wyoming Dinosaur Center)

Fig. IX–5.34: Horn-Faced Dinosaur (*Albertaceratops*) Skeletal Reconstruction from Alberta (WDC)

Fig. IX–5.35: Horn-Faced Dinosaur (*Albertaceratops*) Skeletal Reconstruction from Alberta (WDC)

Fig. IX–5.36: Another *Triceratops* Skull Reconstruction (WDC)

Fig. IX–5.37: *Triceratops* Skeletal Reconstruction at Houston Museum of Natural Science

Fig. IX–5.38: *Triceratops* Fossilized Skin Section at Houston Museum

Fig. IX–5.39: Close up Photo of Triceratops Fossil Skin

Fig. IX–5.40: Perhaps Hair Bristles Grew out of Circular Holes in *Triceratops* Skin?

Dinosaur Data by Created Kind

Fig. IX–5.41: *Triceratops* Head Connection to Spine at Houston Museum

Fig. IX–5.42: Top View of *Triceratops* Skeletal Reconstruction at Houston Museum

Dinosaur Classification Form

CREATED KIND #6 – HORN-NOSED BIPEDAL DINOSAURS

1. **Created Kind:** Horn-Nosed Bipedal Dinosaurs
2. **Representative Genus Name and Meaning:** *Ceratosaurus* (seh-RAT-oh-SAW-rus). Horned lizard (Greek).
3. **Day Created:** Six.
4. **Hip Design:** Lizard-Hipped (Saurischian).
5. **Teeth Design:** Large and curved toward back of mouth.
6. **Skull Design:** Large with a robust lower jaw.
7. **Feet Design(s):** Three long forward-facing clawed toes, and a four-fingered hand.
8. **Stance Design:** Bipedal.
9. **Size:** Up to 20 feet long.
10. **States (Provinces) Where Fossils Have Been Found:** CO, UT
11. **When First Discovered, Where, and by Whom:** Described by O.C. Marsh in 1884 and discovered in 1883 in Colorado.
12. **Extent of Fossils Found:** Two partial skulls, two nearly complete skeletons and assorted miscellaneous fragments.
13. **Synonyms (Other likely North American genera for the same dinosaur kind):** None.
14. **Other Interesting Facts:** This dinosaur is described as being quite similar to the tyrant dinosaurs except it had a prominent bony horn on its snout and had four fingers on each hand instead of three as with most of the tyrants. Some scientists report that this dinosaur had a line of bony plates that ran down its back and it is thought by others that its tail was more flexible than the tyrant dinosaurs.

Marsh's 1894 holotype specimen is located at the Smithsonian in Washington D.C. and is identified with catalogue No. 4735. The fossil material consists of a "fragmentary skull" and is described at the museum website as follows: "For over 125 years this one specimen has been the most representative member of this genus and has served as the sole basis for all subsequent reconstructions."

The other most referenced *Ceratosaurus* specimen is catalog number 021706 at the Carnegie Museum of Natural History in Pittsburgh. It is described as a "partial jaw."

The above North American natural history museum displayed materials are in accord with the described fossil material findings, i.e. very skimpy in both quality and quantity. I wonder if the nose "horn" on *Ceratosaurus* could be the result of injury, disease or genetic defect.

If so, the *Ceratosaurus* could be moved to the Tyrant Bipedal created kind (#15) and dinosaur created kind #6 would no longer be needed.

15. **Artistic Reconstruction:**

Fig. IX–6: Kind #6: Horn-Nose Bipedal Dinosaur Artistic Reconstruction by M. Pike

16. **Photographs:**

Fig. IX–6.1: *Ceratosaurus* Holotype Skull Cast #4735 with Nose Horn (Smithsonian Institute)

Fig. IX–6.2: *Ceratosaurus* Partial Skeletal Reconstruction
(Smithsonian Institute)

Dinosaur Classification

CREATED KIND #7 – KILLER-CLAWED BIPEDAL DINOSAURS

1. **Created Kind:** Killer-Clawed Bipedal Dinosaurs
2. **Representative Genus Name and Meaning:** *Deinonychus* (die-NON-ee-kus). Terrible claw (Greek).
3. **Day Created:** Six.
4. **Hip Design:** Lizard-Hipped (Saurischian)
5. **Teeth Design:** Large, curved and bladelike with serrated edges.
6. **Skull Design:** Head was large compared to the body, with large eyes.
7. **Feet Design(s):** Four toes on the foot (second toe was the large 'killer' claw) and it had three fingers with claws.
8. **Stance Design:** Bipedal.
9. **Size:** Up to 20 feet long.
10. **States (Provinces) Where Fossils Have Been Found:** AK, MT, NM, OK, TX, UT, WY, (AB)
11. **When First Discovered, Where, and by Whom:** Grant Meyer and John Ostrom first discovered *Deinonychus* remains in Montana in 1964. Ostrom described it in 1969. *Troodon* was described from teeth in 1856 by Joseph Leidy. Barnum Brown discovered *Dromaeosaurus* in Alberta in 1914 and described it in 1922. *Utahraptor* was not described until 1993 from a skull found (you guessed it) in Utah.
12. **Extent of Fossils Found:** More than eight articulated and disarticulated skeletons and skulls.
13. **Synonyms (Other likely North American genera for the same dinosaur kind):**

    Dromaeosaurus (droh-may-oh-SAW-rus). Running lizard (Greek).

    Sauronitholestes (SAWR-ORN-ith-oh-LESS-tees). Lizard bird robber (Greek).

    Troodon (TROO-oh-don). Wounding tooth (Greek).

    Utahraptor (YOU-tah-RAP-tor). Utah robber (English/Latin).

14. **Other Interesting Facts:** The sickle-shaped "killer" or "terrible" claw is so unusual that it looks unnatural in reconstructions of the complete animal. Paleontologists currently believe it must have been held up off the ground while the animal ran along the ground. Probably the best known killer-claw dinosaur is *Velociraptor* (vel-OSS-ee-RAP-tor) whose name in Greek/Latin means 'swift robber.' Its remains were found in Mongolia and it was made famous by its vicious reconstruction in the movie Jurassic Park.

15. Artistic Reconstruction:

Fig. IX–7: Kind #7: Killer-Clawed Bipedal Dinosaur, Artistic Reconstruction by M. Pike

16. Photographs:

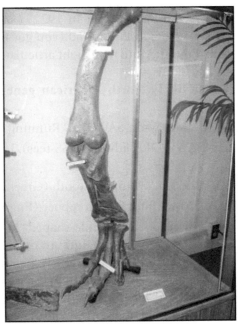

Fig. IX–7.1: *Utahraptor* Leg Reconstruction (BYU Museum Provo, Utah)

Dinosaur Data by Created Kind

Fig. IX–7.2: *Utahraptor* Skeletal Reconstruction (Price, Utah Museum)

Fig. IX–7.3: *Sauronitholestes* Skeletal Reconstruction
(Tyrrell Museum, Alberta)

Fig. IX–7.4: *Deinonychus* Replica Skull from Cloverly Formation, Wyoming (Black Hills Institute)

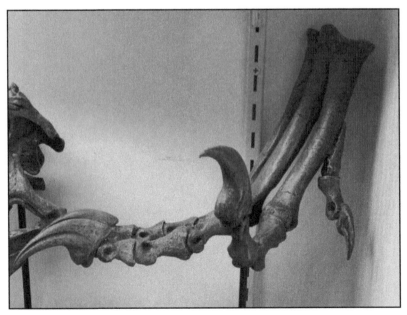

Fig. IX–7.5: *Deinonychus* Foot with Killer Claw from Cloverly Formation, Wyoming (Black Hills Institute)

Dinosaur Data by Created Kind 141

Fig. IX–7.6: *Dromaeosaurus* Replica Skull Tyrrell 84.8.1 from Judith River Fm., Alberta (Black Hills Institute)

Dinosaur Classification

CREATED KIND #8 – LITHE, FAST RUNNING DINOSAURS

1. **Created Kind:** Lithe, Fast Running Dinosaurs
2. **Representative Genus Name and Meaning:** *Coelophysis* (SEE-loe-FIE-sis). Hollow nature (Greek).
3. **Day Created:** Six.
4. **Hip Design:** Lizard-Hipped (Saurischian).
5. **Teeth Design:** Sharp backward curved and designed for slashing. Upper teeth were longer than the lower.
6. **Skull Design:** Long and lightly built with several 'windows.'
7. **Feet Design(s):** Three clawed toes on feet, three clawed fingers on hands.
8. **Stance Design:** Bipedal.
9. **Size:** Up to ten feet long.
10. **States (Provinces) Where Fossils Have Been Found:** AZ, MA, NM, UT.
11. **When First Discovered, Where, and by Whom:** *Coelophysis* was first described by E.D. Cope in 1889 from fragments found in New Mexico. Dozens of skeletons of *Coelophysis* were found at Ghost Ranch in New Mexico in 1947 by Edwin Colbert of the American Museum of Natural History.
12. **Extent of Fossils Found:** Several hundred individual *Coelophysis* specimens have been recovered in New Mexico including nearly complete articulated skeletons. The extent of fossil finds for the other genera are mostly fragmentary.
13. **Synonyms (Other likely North American genera for the same dinosaur kind):** *Camposaurus* (KAMP-oh-SAW-rus). From Charles Lewis Camp (English/Greek).
    Chindesaurus (CHIN-dee-SAW-rus). Chinde Point lizard (English/Greek).
    Coelurus (KOH-eh-LURE-us). Hollow wild ox (Greek).
    Dilophosaurus (die-LOH-foh-SAW-rus). Double-crested lizard (Greek).
    Eucoelophysis (YOU-SEE-loe-FIE-sis). True *Coelophysis* (Greek).
    Gojirasaurus (GO-JYE-ruh-SAW-rus). Gojira lizard (Japanese/Greek).
    Ornitholestes (ORN-ith-oh-LESS-tees). Bird robber (Greek).
    Podokesaurus (POD-oh-KEE-SAW-rus). Swift foot lizard (Greek).
    Segisaurus (SEG-ee-SAW-rus). Segi Canyon lizard (English/Greek).

14. **Other Interesting Facts:** These animals were slimly built with a lightweight body and a long tail. These are characteristics that indicate they could run very fast. Most secular paleontologists believe the large quantity of *Coelophysis* remains are at Ghost Ranch as the result of a large flood.
15. **Artistic Reconstruction:**

Fig. IX–8: Kind #8: Lithe, Fast Running Dinosaur
Artistic Reconstruction by M. Pike

16. **Photographs:**

Fig. IX–8.1: *Coelophysis* Replica Skeleton AMNH 7223 from Kayenta Fm., New Mexico (Black Hills Institute)

144　　　　　*Guidebook to North American Dinosaurs*

Fig. IX–8.2: *Coelophysis* Skeleton from New Mexico
(Wyoming Dinosaur Center)

Fig. IX–8.3: Close-up of *Coelophysis* Skull from New Mexico
(Wyoming Dinosaur Center)

Dinosaur Classification
CREATED KIND #9 – LONG-NECKED BIG-CLAWED DINOSAURS

1. **Created Kind:** Long-Necked Big-Clawed Dinosaurs
2. **Representative Genus Name and Meaning:** *Anchisaurus* (an-kee-SAW-rus). Near lizard (Greek).
3. **Day Created:** Six.
4. **Hip Design:** Lizard-Hipped (Saurischian).
5. **Teeth Design:** Coarsely-serrated and pencil-shaped.
6. **Skull Design:** Small and lightly built with a slender snout.
7. **Feet Design(s):** Four-toed rear foot. Five-fingered hand with the first finger consisting of a much enlarged, sharply-curved claw.
8. **Stance Design:** Mostly bipedal, but probably browsed on all fours.
9. **Size:** Up to about eight feet long.
10. **States (Provinces) Where Fossils Have Been Found:** CN, MA
11. **When First Discovered, Where, and by Whom:** *Anchisaurus* was first discovered in 1818 but was recognized as a dinosaur in 1885 by O.C. Marsh.
12. **Extent of Fossils Found:** Only about five incomplete skeletons and skulls have been found in North America.
13. **Synonyms (Other likely North American genera for the same dinosaur kind):** *Ammosaurus* (am-moh-SAW-rus). Knot lizard (Greek).
    Plateosaurus (PLAY-tee-oh-SAW-rus). Flat lizard (Greek) [From Europe]
14. **Other Interesting Facts:** Paleontologist Robert T. Bakker wrote, "*Anchisaurus* displayed no body armor, but wielded huge curved claws on their powerfully muscled thumbs and long pointed claws on their stout hind feet." He speculated that *Anchisaurus* defended against tyrant dinosaur attacks similarly to the way modern anteaters and cassowary birds fight—slashing out from an erect two-footed stance.

    Other secular paleontologists have written that they believe *Anchisaurus* is closely related to the dinosaur *Plateosaurus* whose remains have been found in relatively great abundance in Germany, Switzerland, and France. Much of what has been written in the secular literature about the North American dinosaur relies on the *Plateosaurus* material. An inspection of the available fossils from the two genera reinforces that opinion—in fact they certainly look as if they could be the same created kind. The evidence that the North American *Anchisaurus* fossils have been found at the same 45° position of latitude (as well as adjacent to the Atlantic Ocean) as the European *Plateosaurus* fossils would match the expectations of creationist models for the worldwide flood.

Since I have not run into any skeletal reconstructions of *Anchisaurus* in North American museums, the photos that follow of the Long-Necked Big-Clawed dinosaurs are of *Plateosaurus* displays from Europe.

15. **Artistic Reconstruction:**

Fig. IX–9: Kind #9: Long-Necked Big-Clawed Dinosaurs, Artistic Reconstruction by M. Pike

Dinosaur Data by Created Kind

16. **Photographs:**

Fig. IX–9.1: *Plateosaurus* Skeletal Reconstruction from Europe

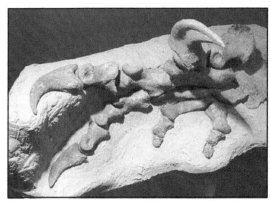

Fig. IX–9.2: Reconstruction of Hand of *Plateosaurus* (Europe)

Fig. IX–9.3: European Reconstruction of *Plateosaurus*

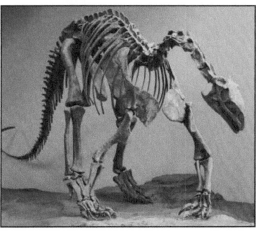

Fig. IX–9.4: Reconstruction showing Head, Neck, Arms and Hands of *Plateosaurus* (Europe)

Dinosaur Classification

CREATED KIND #10 – LONG-NECKED BOXY-HEADED DINOSAURS

1. **Created Kind:** Long-Necked Boxy-Headed Dinosaurs
2. **Representative Genus Name and Meaning:** *Brachiosaurus* (BRAK-ee-oh-SAW-rus). Arm lizard (Greek).
3. **Day Created:** Six.
4. **Hip Design:** Lizard-Hipped (Saurischian).
5. **Teeth Design:** Chisel-shaped upper and lower teeth distributed all around the jaws.
6. **Skull Design:** Boxy with large holes for the nostrils and eye sockets.
7. **Feet Design(s):** Five widely splayed toes, with the inner toe, longer and having a curved claw.
8. **Stance Design:** Quadrupedal.
9. **Size:** Up to 75 feet long.
10. **States (Provinces) Where Fossils Have Been Found:** CO, MT, NM, UT, WY, TX.
11. **When First Discovered, Where, and by Whom:** Genus *Brachiosaurus* was first discovered in 1900 in Utah by Elmer G. Riggs. *Camarasaurus* was first described by E.D. Cope in 1877.
12. **Extent of Fossils Found:** The fossils for these long-necked boxy-heads are the most common of North American Sauropods with over a dozen complete or partial skeletons of *Camarasaurus,* almost as many skulls of *Camarasaurus* and two partial skeletons of *Brachiosaurus* found in the United States. Only one side of part of the skeleton of *Cedarosaurus* is known. The reported fossil finds for *Sauroposeidon* consist of just four huge cervical vertebrae.
13. **Synonyms (Other likely North American genera for the same dinosaur kind):** *Camarasaurus* (KAM-uh-ruh-SAW-rus). Vaulted chamber lizard (Greek).

    Cedarosaurus (see-DAR-oh-SAW-rus). Lizard from Cedar Mountain (Greek).

    Paluxysaurus (puh-LUX-ee-SAW-rus). Lizard from Paluxy (Greek).

    Sauroposeidon (SAW-roh-po-SIE-dun). Lizard of Poseidon (Greek).
14. **Other Interesting Facts:** At one time paleontologists thought that the nostrils on top of the head and the long necks indicated that these animals lived in the water. Current opinions are that they lived on land, using their long necks for browsing in the treetops in a manner

similar to modern giraffes. Teeth design and gastroliths found with skeletal fossils suggest these animals had stones in their gizzards to help digestion.

15. **Artistic Reconstruction:**

Fig. IX–10: Kind #10: Long-Necked Boxy-Headed Dinosaur
Artistic Reconstruction by M. Pike

16. **Photographs:**

Fig. IX–10.1: *Camarasaurus* Skeletal Reconstruction
(Carnegie Museum, Pittsburgh)

Dinosaur Data by Created Kind

Fig. IX–10.2: Comparison of Slender-Headed (L) to Boxy-Headed (R) Long-Necked Dino Skulls (Smithsonian)

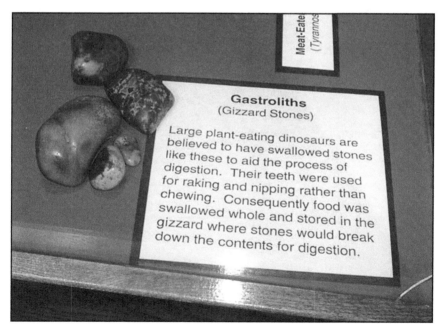

Fig. IX–10.3: Gastroliths or Gizzard Stones (Price, Utah Museum)

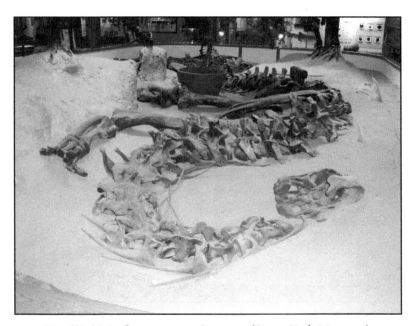

Fig. IX–10.4: *Camarasaurus* Remains (Price, Utah Museum)

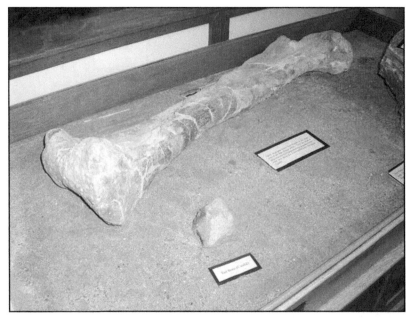

Fig. IX–10.5: Long-Necked Dinosaur Femur Bone (Price, Utah Museum)

Dinosaur Data by Created Kind

Fig. IX–10.6: *Brachiosaurus* Femur (BYU Museum Provo, Utah)

Fig. IX–10.7: *Brachiosaurus* Scapula (BYU Museum, Provo, Utah)

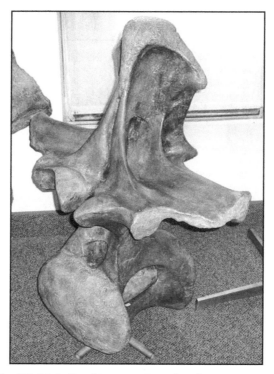

Fig. IX–10.8: *Brachiosaurus* Vertebra (BYU Museum)

Fig. IX–10.9: *Brachiosaurus* Humerus (BYU Museum Provo, Utah)

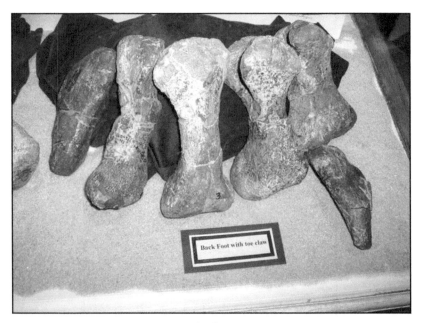

Fig. IX–10.10: Long-Necked Dinosaur Rear Foot Bones
(Price, Utah Museum)

Dinosaur Data by Created Kind

Fig. IX–10.11: Long-Necked Dinosaur Front Foot Bones (Price, Utah Museum)

Fig. IX–10.12: *Camarasaurus* Leg Bones (Cleveland-Lloyd Quarry)

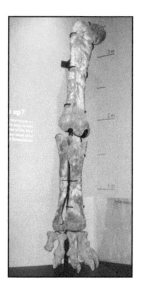

Fig. IX–10.13: *Camarasaurus* Leg Bones (Tyrrell Museum, Alberta)

Fig. IX–10.14: *Camarasaurus* Replica Skull from Morrison Formation, Wyoming (Black Hills Institute)

Fig. IX–10.15: *Camarasaurus* Replica Skull from Morrison Fm., WY (Black Hills Institute) a.k.a. "E.T."

Dinosaur Data by Created Kind

Fig. IX–10.16: *Camarasaurus* Skull and Femur from Morrison Fm., WY (Black Hills Institute) a.k.a. "Elaine"

Fig. IX–10.17: *Camarasaurus* Tail Vertebra
(Wyoming Dinosaur Center)

Fig. IX–10.18: *Paluxysaurus* Skeletal Reconstruction at Ft. Worth Museum

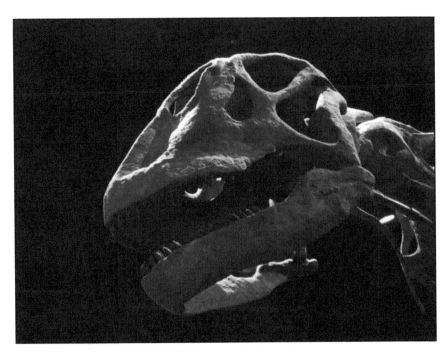

Fig. IX–10.19: *Paluxysaurus* Skull on Reconstruction at Ft. Worth Museum

Dinosaur Data by Created Kind

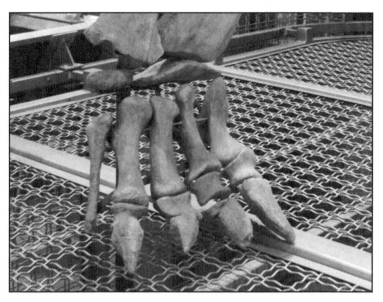

Fig. IX–10.20: Rear Foot on *Paluxysaurus* Reconstruction at Ft. Worth Museum

Fig. IX–10.21: Body of *Paluxysaurus* Reconstruction at Ft. Worth Museum

Fig. IX–10.22: *Paluxysaurus* Pelvic Girdle at Ft. Worth Museum

Fig. IX–10.23: *Paluxysaurus* Reconstructed Skeleton Front View at Ft. Worth Museum

Dinosaur Classification

CREATED KIND #11 – LONG-NECKED SLENDER-HEADED DINOSAURS

1. **Created Kind**: Long-Necked Slender-Headed Dinosaurs
2. **Representative Genus Name and Meaning**: *Diplodocus* (dip-LOH-doh-kus). Double beam (Greek).
3. **Day Created**: Six.
4. **Hip Design**: Lizard-Hipped (Saurischian).
5. **Teeth Design**: Pencil-like teeth arranged like a rake in front of long jaws. No teeth at the back of mouth.
6. **Skull Design**: Slender.
7. **Feet Design(s)**: Five splayed toes.
8. **Stance Design**: Quadrupedal.
9. **Size**: Up to 90 feet long.
10. **States (Provinces) Where Fossils Have Been Found**: CO, NM, OK, UT, WY
11. **When First Discovered, Where, and by Whom**: First discovered and named by O.C. Marsh from remains found in 1878 in Colorado.
12. **Extent of Fossils Found**: At least 20 partial skeletons have been found in the United States, most missing the skulls. About half a dozen skulls have been found so far.
13. **Synonyms (Other likely North American genera for the same dinosaur kind)**: *Alamosaurus* (al-uh-MO-SAW-rus). Lizard from the Ojo Alamo sandstone (Spanish/Greek).
    Apatosaurus (uh-PAT-oh-SAW-rus). Deceptive lizard (Greek).
    Barosaurus (BAH-roh-SAW-rus). Heavy lizard (Greek).
    Eobrontosaurus (EE-oh-BRONT-oh-SAW-rus). Dawn thunder lizard (Greek).
    Seismosaurus (SIZE-moh-SAW-rus). Earthquake lizard (Greek).
    Supersaurus (SOUP-er-SAW-rus). Super lizard (Greek).
14. **Other Interesting Facts**: Teeth design suggests that *Diplodocus* may have used "gizzard" stones to grind up its food. Due to the lack of discovered fossil skulls the dinosaur named *Brontosaurus* was fitted with an incorrect skull when first reconstructed for display. Therefore *Brontosaurus* was later superseded by the genus name *Apatosaurus* that had been assigned earlier than *Brontosaurus* for the correct reconstruction. *Supersaurus* genus is thought by some paleontologists to have grown to nearly one hundred feet long, but that is based on

incomplete fossil evidence. *Diplodocus* is the longest dinosaur known from a complete skeleton.

15. **Artistic Reconstruction:**

Fig. IX–11: Kind #11: Long-Necked Slender-Headed Dinosaur, Artistic Reconstruction by M. Pike

16. **Photographs:**

Fig. IX–11.1: *Diplodocus* Skeletal Reconstruction (Utah Field House Museum)

Dinosaur Data by Created Kind 163

Fig. IX–11.2: Long-Necked Slender-Headed Dinosaur Skeletal Reconstruction (Tyrrell Museum, AB)

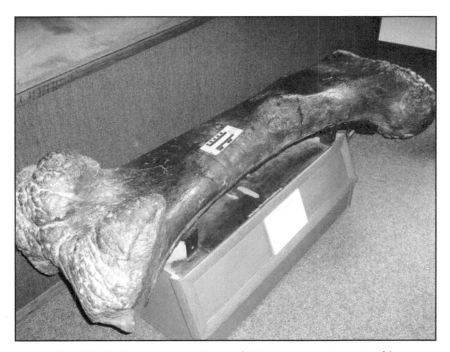

Fig. IX–11.3: *Apatosaurus* Femur (BYU Museum, Provo, Utah)

Fig. IX–11.4: *Diplodocus* Skull (Price, Utah Museum)

Fig. IX–11.5: *Apatosaurus* Skull (Price, Utah Museum)

Dinosaur Data by Created Kind

Fig. IX–11.6: *Apatosaurus* Vertebrae (BYU Museum in Provo, Utah)

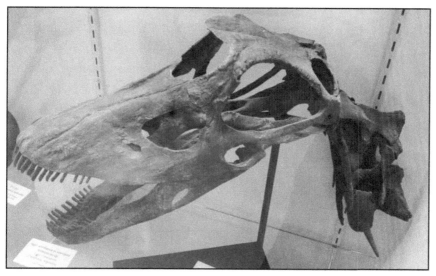

Fig. IX–11.7: *Diplodocus* Replica Skull from Morrison Fm., Wyoming (Black Hills Institute)

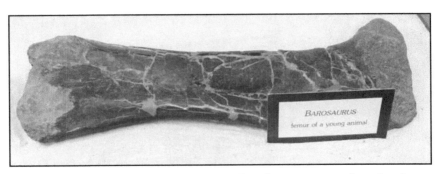

Fig. IX–11.8: *Barosaurus* Femur (South Dakota Museum of Geology)

Fig. IX–11.9: *Diplodocus* Digit (Wyoming Dinosaur Center, Thermopolis, WY)

Dinosaur Data by Created Kind 167

Fig. IX–11.10: *Diplodocus* pes (hind foot) fossil from Wyoming
(Wyoming Dinosaur Center)

Fig. IX–11.11: *Apatosaurus* Metatarsal from Wyoming
(Wyoming Dinosaur Center)

Fig. IX–11.12: *Diplodocus* Dorsal Vertebrae from Wyoming (WDC)

Fig. IX–11.13: *Supersaurus* Skeletal Reconstruction from WY at Wyoming Dinosaur Center in Thermopolis, WY

Fig. IX–11.14: *Supersaurus* Cervical (Neck) Vertebra from WY (Wyoming Dinosaur Center)

Fig. IX–11.15: *Supersaurus* Cervical (Neck) Vertebra from WY (Wyoming Dinosaur Center)

Fig. IX–11.16: *Supersaurus* Dorsal Vertebra from WY
(Wyoming Dinosaur Center)

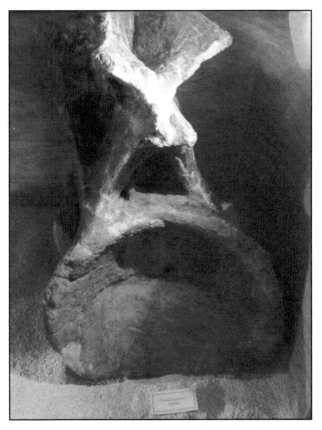

Fig. IX–11.17: *Supersaurus* Dorsal Vertebra from WY
(Wyoming Dinosaur Center)

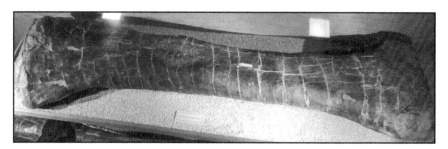

Fig. IX–11.18: *Supersaurus* Fibula Bone from WY
(Wyoming Dinosaur Center)

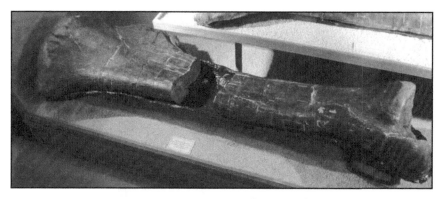

Fig. IX–11.19: *Supersaurus* Tibia Bone from WY
(Wyoming Dinosaur Center)

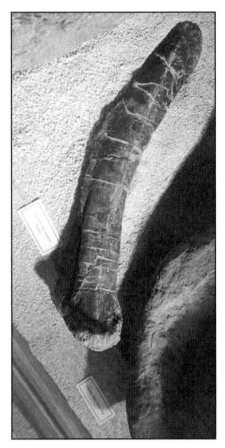

Fig. IX–11.20: *Supersaurus* Chevron Bone from WY (Wyoming Dinosaur Center)

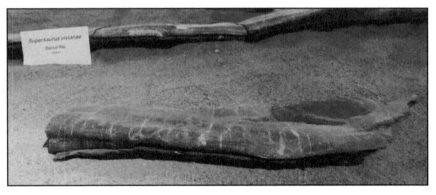

Fig. IX–11.21: *Supersaurus* Ribs from WY (Wyoming Dinosaur Center)

Fig. IX–11.22: *Supersaurus* Vertebrae from WY (Wyoming Dinosaur Center)

Fig. IX–11.23: *Supersaurus* Caudal (Tail) Vertebrae from WY
(Wyoming Dinosaur Center)

Fig. IX–11.24: *Barosaurus* Skeletal Reconstruction at the American Museum of Natural History

Fig. IX–11.25: *Diplodocus* Skeletal Reconstruction at Houston Museum of Natural Science

Dinosaur Classification
CREATED KIND #12 – OSTRICH-LIKE DINOSAURS

1. **Created Kind**: Ostrich-Like Dinosaurs
2. **Representative Genus Name and Meaning**: *Ornithomimus* (ORE-nith-oh-MY-mus). Bird mimic (Greek).
3. **Day Created**: Six.
4. **Hip Design**: Lizard-Hipped (Saurischian).
5. **Teeth Design**: Toothless beak.
6. **Skull Design**: Long and slender with relatively large eyes. Nostrils at the tip of skull.
7. **Feet Design(s)**: Three-toed feet and three-fingered hands, all with quite long claws.
8. **Stance Design**: Bipedal.
9. **Size**: Up to 15 feet long.
10. **States (Provinces) Where Fossils Have Been Found**: CO, UT, WY (AB)
11. **When First Discovered, Where, and by Whom**: *Ornithomimus* was named by O.C. Marsh in 1890. *Struthiomimus* was named by Henry Osborn in 1917 and *Dromiceiomimus* was named by Russell in 1972.
12. **Extent of Fossils Found**: Just four skulls and about twenty incomplete skeletons have been studied so far.
13. **Synonyms (Other likely North American genera for the same dinosaur kind)**: *Dromiceiomimus* (DROM-ee-see-oh-MY-mus). Emu mimic (Greek).
    Struthiomimus (STROO-thee-oh-MY-mus). Ostrich mimic (Latin/Greek).
14. **Other Interesting Facts**: These dinosaurs resembled ostrich-like running birds in the structure of their heads, necks and legs. They were different from ostriches in that they had long forelimbs and no wings. Gastroliths (gizzard stones) have been found at the front of the fossil rib cage in a *Struthiomimus* find. One paleontologist suggests these dinosaurs gave birth to live young rather than laying eggs.

Dinosaur Data by Created Kind 177

15. Artistic Reconstruction:

Fig. IX–12: Kind #12: Ostrich-Like Dinosaur, Artistic Reconstruction by M. Pike

16. Photographs:

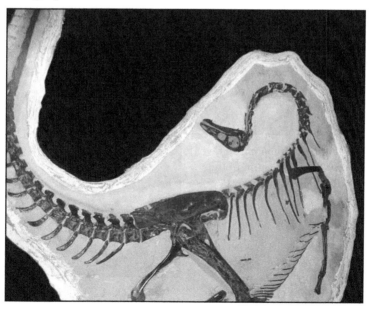

Fig. IX–12.1: *Ornithomimus* Skeleton (Tyrrell Museum, Alberta)

Fig. IX–12.2: *Ornithomimus* Skeletal Reconstructions (Tyrrell Museum, Alberta)

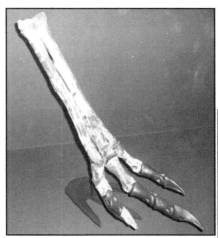

Fig. IX–12.3: *Ornithomimus* Lower Leg and Foot (Museum of the Rockies, Montana)

Fig. IX–12.4: *Ornithomimus* Skeletal Parts Display (Museum of the Rockies)

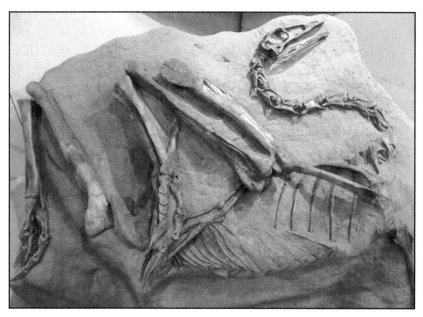

Fig. IX–12.5: *Struthiomimus* Skeleton from Lance Formation WY (Black Hills Institute, SD)

Fig. IX–12.6: *Close*-up of *Struthiomimus* Skull from Lance Formation WY (Black Hills Institute, SD)

Dinosaur Classification

CREATED KIND #13 – PLATE-BACKED DINOSAURS

1. **Created Kind:** Plate-Backed Dinosaurs
2. **Representative Genus Name and Meaning:** *Stegosaurus* (STEG-oh-SAW-rus). Roof lizard (Greek).
3. **Day Created:** Six.
4. **Hip Design:** Bird-Hipped (Ornithischian).
5. **Teeth Design:** Toothless premaxilla with beak and leaf-shaped serrated back teeth.
6. **Skull Design:** Slender with small brain cavity.
7. **Feet Design(s):** Five toes on front feet and four toes on the rear feet. All feet hoof-like.
8. **Stance Design:** Quadrupedal.
9. **Size:** Up to 25 feet long.
10. **States (Provinces) Where Fossils Have Been Found:** CO, UT, WY.
11. **When First Discovered, Where, and by Whom:** First described by O.C. Marsh in 1877. A mostly complete skeleton of *Stegosaurus* was found in 1886 in Colorado by Marsh's crew and was reconstructed and placed on display in the Smithsonian not long after.
12. **Extent of Fossils Found:** Stegosaurs are relatively rare with only three complete fossil skeletons and two partial skeletons reported found in United States sediments.
13. **Synonyms (Other likely North American genera for the same dinosaur kind):** *Hesperosaurus* (HES-purr-oh-SAW-rus). Western lizard (Greek).
14. **Other Interesting Facts:** The plate-backed dinosaurs had relatively small heads and brains, however current expert opinion is that was not necessarily an indication of stupidity. The plates on the back are thought to have had the purpose of defense, sexual display and/or heat regulation. They had long spikes on their tails that could have been used as defensive weapons. The front legs were considerably shorter than the rear legs, a design that would force the head down toward ground level food, however some think these animals could rear up on their hind legs to eat. The teeth design suggests that they may have used gizzard stones to grind up food in their digestive tract.

15. Artistic Reconstruction:

Fig. IX–13: Kind #13: Plate-Backed Dinosaur, Artistic Reconstruction by M. Pike

16. Photographs:

Fig. IX–13.1: *Stegosaurus* Skeletal Reconstruction (Carnegie Museum, Pittsburgh)

Fig. IX–13.2: *Stegosaurus* Skeletal Reconstruction
(Utah Field House Museum)

Fig. IX–13.3: *Stegosaurus* Head and Front of Skeleton
(Cleveland-Lloyd Quarry)

Dinosaur Data by Created Kind 183

Fig. IX–13.4: *Stegosaurus* Remains (Price, Utah Museum)

Fig. IX–13.5: *Stegosaurus* Bony Plate (Price, Utah Museum)

Fig. IX–13.6: *Stegosaurus* Tail Spike (Price, Utah Museum)

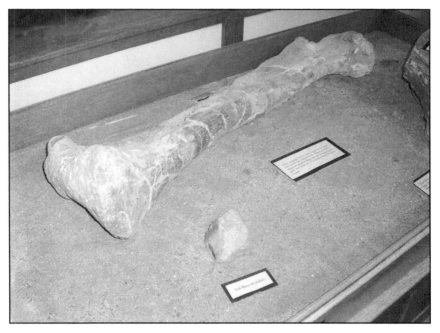

Fig. IX–13.7: *Stegosaurus* Bones (Price, Utah Museum)

Dinosaur Data by Created Kind 185

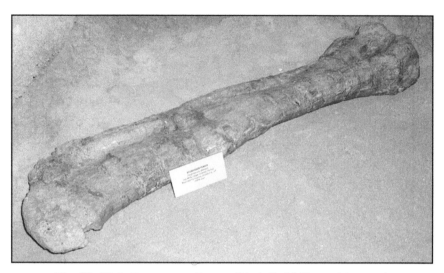

Fig. IX–13.8: *Stegosaurus* Femur (Utah Field House Museum)

Fig. IX–13.9: *Stegosaurus* Remains (BYU Museum Provo, Utah)

Fig. IX–13.10: *Stegosaurus* Armor Plate (BYU Museum Provo, Utah)

Fig. IX–13.11: *Hesperosaurus* Skeletal Reconstruction at Houston Museum of Natural Science

Fig. IX–13.12: *Hesperosaurus* Skull on Reconstruction at Houston Museum

Fig. IX–13.13: *Hesperosaurus* Tail Spikes on Reconstruction at Houston Museum

Fig. IX–13.14: *Hesperosaurus* Bony Back Plates on Reconstruction at Houston Museum

Fig. IX–13.15: *Stegosaurus* "Active" Skeletal Reconstruction at Houston Museum

Dinosaur Classification
CREATED KIND #14 – THICK-HEADED DINOSAURS

1. **Created Kind**: Thick-Headed Dinosaurs
2. **Representative Genus Name and Meaning**: *Pachycephalosaurus* (PAK-ee-kef-AH-loh-SAW-rus). Thick-headed lizard (Greek).
3. **Day Created**: Six.
4. **Hip Design**: Bird-Hipped (Ornithischian).
5. **Teeth Design**: Triangular-shaped with serrations.
6. **Skull Design**: Dome-shaped with a thick bony top.
7. **Feet Design(s)**: Three or four toes on each foot.
8. **Stance Design**: Bipedal.
9. **Size**: Up to 25 feet long?
10. **States (Provinces) Where Fossils Have Been Found**: MT, ND, SD, WY, (AB)
11. **When First Discovered, Where, and by Whom**: Lawrence Lambe is credited with naming *Stegoceras* fossils from Alberta in 1902. Gilmore named a nearly complete thick-headed dinosaur skull from Wyoming in 1931 *Troodon wyomingensis,* but that was renamed in 1943 by Brown and Schlaikjer to genus *Pachycephalosaurus*.
12. **Extent of Fossils Found**: In North America only two complete skulls, two partial skulls, about twenty skull fragments, and a few skeletal parts have been reported found. Reconstructions of the bodies of these animals are primarily based on fossils of animals thought to be similar found in Mongolia and elsewhere in the world.
13. **Synonyms (Other likely North American genera for the same dinosaur kind)**: *Sphaerotholus* (SPHERE-oth-uh-lus). Spherical dome (Greek).
    Stegoceras (STEG-oh-SEE-rus). Roof horn (Greek).
    Stygimoloch (STY-gee-MOH-lok). Lower-world idol (Greek/Hebrew).
14. **Other Interesting Facts**: The purpose for the thickened domed skull roofs of these animals is unknown. Common speculations are that the domes were used for head-to-head butting, flank butting, or sexual display. Variations in the domes from specimen to specimen could indicate differences between males and females.

15. Artistic Reconstruction:

Fig. IX–14: Kind #14: Thick-Headed Dinosaur, Artistic Reconstruction by M. Pike

16. Photographs:

Fig. IX–14.1: Thick-Headed Dinosaur Skulls (Museum of the Rockies, Montana)

Dinosaur Data by Created Kind 191

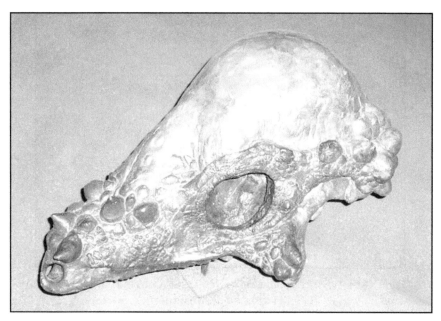

Fig. IX–14.2: *Pachycephalosaurus* Skull (Tyrrell Museum, Alberta)

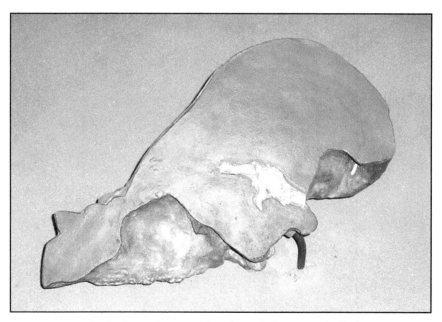

Fig. IX–14.3: Cross-Section of *Pachycephalosaurus* Skull Showing Bone and Brain Cast (Tyrrell)

Fig. IX–14.4: *Pachycephalosaurus* Replica Skull from Hell Creek Fm., MT (Black Hills Institute)

Fig. IX–14.5: Thick-Headed Replica Skull from Hell Creek Fm. SD (Black Hills Institute)

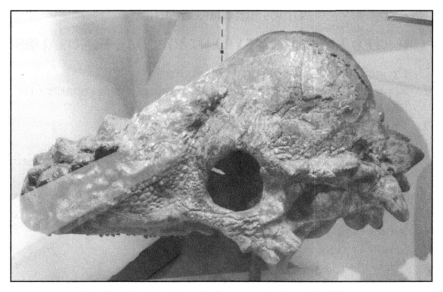

Fig. IX–14.6: *Pachycephalosaurus* Replica Skull from Lance Formation, Wyoming (BHI)

Fig. IX–14.7: *Pachycephalosaurus* Skull Fragment (South Dakota Museum of Geology)

Dinosaur Classification

CREATED KIND #15 – TYRANT BIPEDAL DINOSAURS

1. **Created Kind**: Tyrant Bipedal Dinosaurs
2. **Representative Genus Name and Meaning**: *Tyrannosaurus* (TIE-RAN-oh-SAW-rus). Tyrant lizard (Latin/Greek).
3. **Day Created**: Six.
4. **Hip Design**: Lizard-Hipped (Saurischian).
5. **Teeth Design**: Large, sharp and curved backwardly teeth. Some teeth have been found that are serrated.
6. **Skull Design**: Large, up to three feet long with window-like openings.
7. **Feet Design(s)**: Four or five toed feet and two or three fingered hands.
8. **Stance Design**: Bipedal.
9. **Size**: Up to 45 feet long.
10. **States (Provinces) Where Fossils Have Been Found**: AK, CO, MT, NM, OK, SD, TX, UT, (AB)
11. **When First Discovered, Where, and by Whom**: A partial skeleton was discovered in Montana in 1902. *Tyrannosaurus* was described by Henry Osborn in 1905. *Albertosaurus* remains were found by J.B. Tyrell in 1884. *Allosaurus* was named by O.C. Marsh in 1877.
12. **Extent of Fossils Found**: Fossils for at least one hundred individuals have been reported, although most are disarticulated and fragmentary. The *Tyrannosaurus* named "Sue" on display in Chicago is over 90% complete. As many as 44 individual *Allosaurus* are thought to be represented by the disarticulated fossils taken from the Cleveland-Lloyd quarry in Utah.
13. **Synonyms (Other likely North American genera for the same dinosaur kind)**: *Acrocanthosaurus* (AK-roh-CAN-thoh-SAW-rus). High-spined lizard (Greek).
    Albertosaurus (al-BERT-oh-SAW-rus). Alberta lizard (English/Greek)
    Allosaurus (al-oh-SAW-rus). Different lizard (Greek).
    Appalachiosaurus (AP-uh-LATCH-ee-oh-SAW-rus). Lizard of the Appalachians (English/Greek).
    Daspletosaurus (DAS-plee-toh-SAW-rus). Frightful lizard (Greek).
    Edmarka (ED-mark-uh). Named after William Edmark.
    Gorgosaurus (GORE-go-SAW-rus). Terrible lizard (Greek).

Dinosaur Data by Created Kind 195

Marshosaurus (MARSH-oh-SAW-rus). Marsh's lizard (English/Greek).

Nanotyrannus (NAN-oh-tie-RAN-us). Small tyrant (Latin).

Saurophaganax (SAW-roh-FAG-an-aks). Chief lizard eater (Greek).

Stokesosaurus (STOKE-ee-so-SAW-rus). Stoke's lizard (English/Greek).

Torvosaurus (TORE-voh-SAW-rus). Savage lizard (Latin/Greek).

14. **Other Interesting Facts:** Probably the best known of all dinosaurs due to its ferocious look and very large size. The arms for these dinosaurs were very small compared to the proportions of the rest of the animal. Scientists of today are in disagreement about how fast they ran, what they ate, and how they used their arms. Tooth marks on the fossilized remains of a number of other dinosaurs look like they were made by the tyrant dinosaurs as determined by matching the marks to teeth.

15. **Artistic Reconstruction:**

Fig. IX–15: Kind #15: Tyrant Bipedal Dinosaur, Artistic Reconstruction by M. Pike

16. Photographs:

Fig. IX–15.1: *Allosaurus* Skeletal Reconstruction
(BYU Museum Provo, Utah)

Fig. IX–15.2: *Allosaurus* Skeletal Reconstruction
(Price Utah Museum)

Dinosaur Data by Created Kind 197

Fig. IX–15.3: *Allosaurus* Skeletal Reconstruction (Cleveland-Lloyd Quarry)

Fig. IX–15.4: *Torvosaurus* Skeletal Reconstruction (BYU Museum Provo, Utah)

Fig. IX–15.5: *Tyrannosaurus* Skull (BYU Museum Provo, Utah)

Fig. IX–15.6: *Allosaurus* Leg Bones from Cleveland-Lloyd Quarry in Utah

Dinosaur Data by Created Kind

Fig. IX–15.7: *Allosaurus* Skull (Dinosaur National Monument)

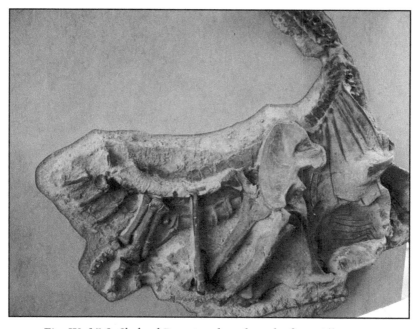

Fig. IX–15.8: Skeletal Remains thought to be from *Allosaurus* (Dinosaur National Monument)

Fig. IX–15.9: *Allosaurus* Forearm and Hand (Dinosaur Nat'l Monument)

Fig. IX–15.10: *Tyrannosaurus* Tooth (Museum of Rockies)

Fig. IX–15.11: *Tyrannosaurus* Broken Femur Where Soft Tissue was Found (Museum of the Rockies)

Dinosaur Data by Created Kind 201

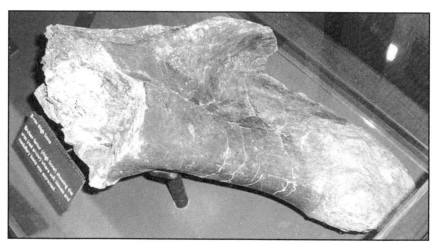

Fig. IX–15.12: Another View of *Tyrannosaurus* Femur
(Museum of Rockies, Montana)

Fig. IX–15.13: *Tyrannosaurus* Skull
(Museum of the Rockies, Montana)

Fig. IX–15.14: *Tyrannosaurus* Skull
(Museum of the Rockies, Montana)

Fig. IX–15.15: *Tyrannosaurus* Ankle and Foot
(Museum of the Rockies Montana)

Fig. IX–15.16: Portion of Actual *Tyrannosaurus* Skeleton Display
(Museum of the Rockies, Montana)

Dinosaur Data by Created Kind

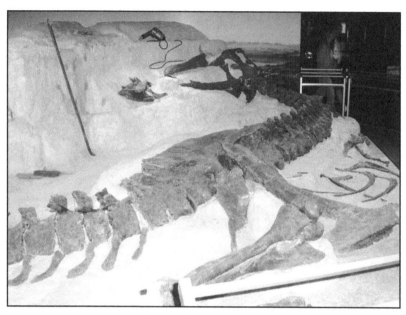

Fig. IX–15.17: More of Actual *Tyrannosaurus* Skeleton on Display (Museum of the Rockies, Montana)

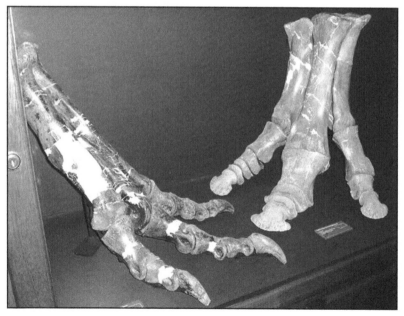

Fig. IX–15.18: Comparison of *Daspletosaurus* Foot (L) to Duck-Billed Foot (R) (Museum of the Rockies, Montana)

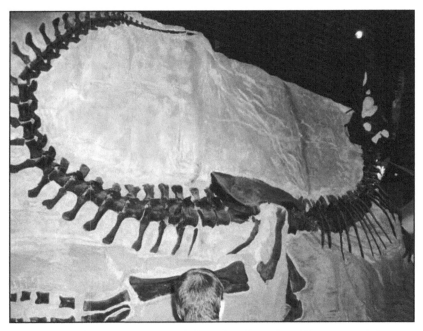

Fig. IX–15.19: *Tyrannosaurus* Skeletal Display (Tyrrell Museum, Alberta)

Fig. IX–15.20: *Gorgosaurus* Skeletal Display (Tyrrell Museum, Alberta)

Dinosaur Data by Created Kind

Fig. IX–15.21: *Tyrannosaurus* Skeletal Reconstruction
(Tyrrell Museum, Alberta)

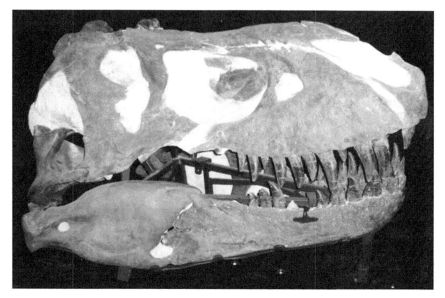

Fig. IX–15.22: *Tyrannosaurus* Skull (Tyrrell Museum, Alberta)

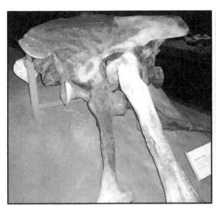

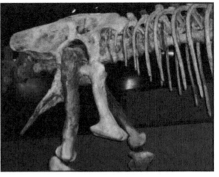

Fig. IX–15.23: Tyrant Bipedal Dinosaur Hip Section (Tyrrell Museum, Alberta)

Fig. IX–15.24: *Tyrannosaurus* Hip Reconstruction (Tyrrell Museum, Alberta)

Fig. IX–15.25: *Allosaurus* Skeletal Reconstruction (San Diego Museum)

Fig. IX–15.26: Tyrant Bipedal Dinosaur Skull (OMSI Portland, OR)

Dinosaur Data by Created Kind

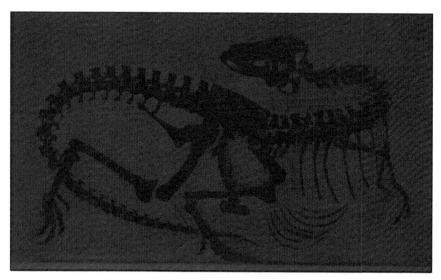

Fig. IX–15.27: *Albertosaurus* Skeletal Display (Smithsonian Washington, DC)

Fig. IX–15.28: *Gorgosaurus* Replica Skull from Two Medicine Fm., MT (Black Hills Institute, SD)

Fig. IX–15.29: *Tyrannosaurus* AMNH-5027 Replica Skull from Montana (Black Hills Institute, SD)

Fig. IX–15.30: *Tyrannosaurus* Replica Skull MOR-008 from Hell Creek Formation, MT (BHI)

Fig. IX–15.31: *Tyrannosaurus* Replica Skull LACM 23844 from Hell Creek Fm., MT (BHI)

Fig. IX–15.32: *Tyrannosaurus* Replica Skull RTMP 81.6.1 from Willow Creek Fm., AB (BHI) a.k.a. "Black Beauty"

Fig. IX–15.33: *Tyrannosaurus* Replica Skull BHI 4100 from Hell Creek Fm., SD (BHI) a.k.a. "Duffy"

Fig. IX–15.34: *Albertosaurus* Replica Skull from Two Medicine Fm., Montana (Black Hills Institute)

Dinosaur Data by Created Kind 211

Fig. IX–15.35: *Albertosaurus* Skeletal Cast Replica Reconstruction Tyrrell 81.10.1 from Judith River Fm., AB (BHI)

Fig. IX–15.36: *Tyrannosaurus* Fossil Bone Skeletal Reconstruction BHI-3033 from Hell Creek Fm., SD (BHI) a.k.a. "Stan"

Fig. IX–15.37: *Nanotyrannus* Replica Skull from Hell Creek Fm., Montana (BHI) a.k.a. "Jane"

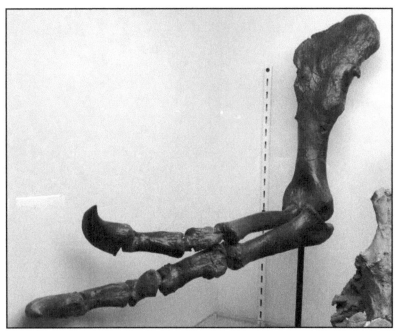

Fig. IX–15.38: *Tyrannosaurus* Replica Cast of Arm of "Sue" from Hell Creek Formation, South Dakota (BHI)

Dinosaur Data by Created Kind 213

Fig. IX–15.39: *Tyrannosaurus* Dentary Bone from Hell Creek Formation, South Dakota (Black Hills Institute)

Fig. IX–15.40: *Tyrannosaurus* Actual Fossil Cranium from Hell Creek Fm., SD (SD Museum of Geology)

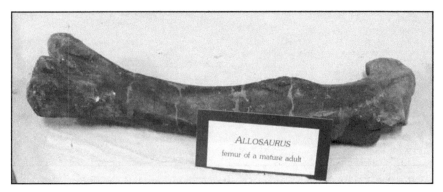

Fig. IX–15.41: *Allosaurus* Femur (South Dakota Museum of Geology)

Fig. IX–15.42: Series of *Allosaurus* Teeth from Wyoming (WDC)

Dinosaur Data by Created Kind

Fig. IX–15.43: *Allosaurus* Replica Skull (Wyoming Dinosaur Center)

Fig. IX–15.44: *Tyrannosaurus* Reconstruction of Right Foot (WDC)

Fig. IX–15.45: *Tyrannosaurus* Reconstruction of Left Foot (WDC)

Fig. IX–15.46: *Tyrannosaurus* "Stan" Skeletal Reconstruction at Houston Museum of Natural Science

Fig. IX–15.47: *Acrocanthosaurus* Skeletal Reconstruction at Houston Museum

Fig. IX–15.48: *Gorgosaurus* Skeletal Reconstruction at Houston Museum

Fig. IX–15.49: Typical Chest Construction of Tyrant Bipedal Dinosaur – See Furcula

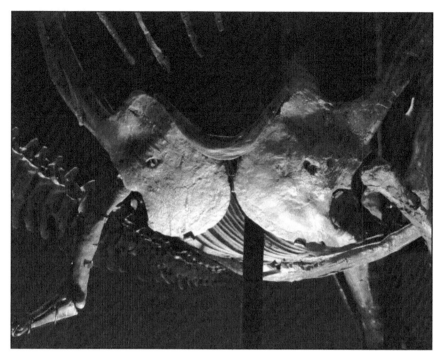

Fig. IX–15.50: Another Tyrant Bipedal Chest with Furcula

Fig. IX–15.51: *Tyrannosaurus* "Bucky" Skeletal
Reconstruction at Houston Museum

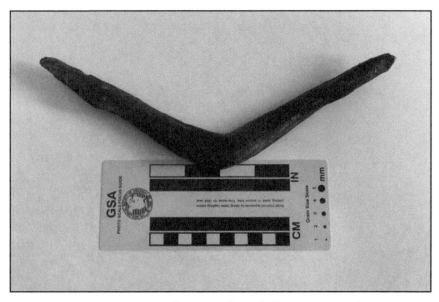

Fig. IX–15.52: Full-Size Replica of *Allosaurus* Furcula

Bibliography

MUCH OF THE information in the *Guidebook to North American Dinosaurs, According to Created Kinds* comes from the author's direct experience in museums and the field. Other information was developed from the following listed references:

Secular Bibliography—Dinosaurs

1. Achenbach, Joel. 2003. "Dinosaurs Come Alive." *National Geographic.* March, p. 2.
2. Alexander, R. McNeill. 1989. *Dynamics of Dinosaurs & Other Extinct Giants.* Columbia University Press, New York.
3. Bakker, Robert. 1986. *The Dinosaur Heresies.* Kensington Publishing, New York.
4. Brett-Surman, Michael, ed. 1997. *A Guide to Dinosaurs.* Fog City Press, San Francisco.
5. Brusatte, Steve. 2009. *Field Guide to Dinosaurs – The Ultimate Dinosaur Encyclopedia.* Quercus Publishing, London, UK.
6. Colbert, Edwin. 1984. *The Great Dinosaur Hunters and Their Discoveries.* Dover Publications, New York.
7. Cranfield, Ingrid, ed. 2002. *The Illustrated Directory of Dinosaurs and other Prehistoric Creatures.* Salamander Books, London.
8. Currie, Philip and Koppelhus, Eva. 1996. *101 Questions about Dinosaurs.* Dover Publications, New York.
9. Currie, Philip and Koppelhus, Eva. 2005. *Dinosaur Provincial Park: A Spectacular Ancient Ecosystem Revealed.* Indiana University Press, Indianapolis, IN.

10. Currie, Philip and Padian, Kevin, eds. 1997. *Encyclopedia of Dinosaurs*. Academic Press, New York.
11. Czerkas, Sylvia and Olson, Everett, eds. 1987. *Volume I Dinosaurs Past and Present*. University of Washington Press, Seattle.
12. Czerkas, Sylvia and Olson, Everett, eds. 1987. *Volume II Dinosaurs Past and Present*. University of Washington Press, Seattle.
13. DeCourten, Frank. 1998. *Dinosaurs of Utah*. University of Utah Press, Salt Lake City.
14. Dixon, Dougal. 2007. *The Complete Book of Dinosaurs*. Anness Publishing, London.
15. Dodson, Peter. 1996. *The Horned Dinosaurs—A Natural History*. Princeton University Press, Princeton, NJ.
16. Edgar, Blake and Gattuso, John, eds. 1999. *Dinosaur Digs*. Discovery Communications, Singapore.
17. Farlow, James O. 1993. *The Dinosaurs of Dinosaur Valley State Park*. Texas Parks and Wildlife Department, Austin, TX.
18. Gillette, J. Lynett. 1997. *Dinosaur Ghosts—The Mystery of Coelophysis*. Scholastic Inc., New York.
19. Glen, William, ed. 1994. *The Mass-Extinction Debates: How Science Works in a Crisis*. Stanford University Press, Stanford, CA.
20. Gore, Rick. 1993. "Dinosaurs." *National Geographic*, Vol. 183 No. 1, January, p. 2.
21. Gross, Renie. 1998. *Dinosaur Country—Unearthing the Alberta Badlands*. Badlands Books, Wardlow, AB.
22. Hilton, Richard. 2003. *Dinosaurs and Other Mesozoic Reptiles of California*. University of California Press, Berkeley, CA.
23. Horner, Jack. 2001. *Dinosaurs under the Big Sky*. Mountain Press Publishing, Missoula, MT.
24. Horner, Jack. 1988. *Digging Dinosaurs: The Search That Unraveled the Mystery of Baby Dinosaurs*. Workman Publishing, New York, NY.
25. Horner, John R. and Lessem, Don. 1993. *The Complete T. Rex*. Simon & Schuster, New York.
26. Jaeger, Edmund. 1955. *A Source-Book of Biological Names and Terms, 3rd Edition*. Charles C. Thomas, Springfield, IL.
27. Jaffe, Mark. 2000. *The Gilded Dinosaur—the Fossil War between E.D. Cope and O.C. Marsh and the Rise of American Science*. Crown Publishers, New York.
28. Larson, Peter and Donnan, Kriston. 2004. *Rex Appeal—the Amazing Story of Sue, the Dinosaur that Changed Science, the Law, and my Life*. Invisible Cities Press, Montpelier, VT.

29. Lockley, Martin and Hunt, Adrian. 1995. *Dinosaur Tracks and other Fossil Footprints of the Western United States*. Columbia University Press, New York.
30. Making, Lisa. 2010. *Inside the Royal Tyrrell Museum of Paleontology*. The Royal Tyrrell Museum Cooperating Society, Drumheller, AB.
31. McGinnis, Helen. 1982. *Carnegie's Dinosaurs*. The Board of Trustees, Carnegie Institute, Pittsburgh.
32. Palmer, Douglas et al eds. 1999. *The Simon & Schuster Encyclopedia of Dinosaurs & Prehistoric Creatures—A Visual Who's Who of Prehistoric Life*. Simon & Schuster, New York, NY.
33. Panczner, William D. 1995. *A Traveler's Guide to Tracking Dinosaurs in the Western United States*. Renaissance House, Frederick, CO.
34. Paul, Gregory. 2010. *The Princeton Field Guide to Dinosaurs*. Princeton University Press, Princeton, NJ.
35. Richardson, Hazel. 2003. *Smithsonian Handbooks—Dinosaurs and Prehistoric Life*. DK Publishing, New York.
36. Romer, Alfred Sherwood. 1966. *Vertebrate Paleontology—Third Edition*. University of Chicago Press, Chicago.
37. Sternberg, Charles H. 1909. *The Life of a Fossil Hunter*. Henry Hunt Company, New York, NY.
38. Svarney, Thomas and Barnes-Svarney, Patricia. 2003. *The Handy Dinosaur Answer Book*. Visible Ink Press, Canton, MI.
39. Wallace, Joseph. 1994. *The American Museum of Natural History's Book of Dinosaurs and other Ancient Creatures*. Simon & Schuster, New York.
40. Weishampel, David et al eds. 2007. *Dinosauria—Second Edition*. University of California Press, Berkeley, CA.
41. West, Linda and Chure, Dan. 2001. *Dinosaur—The Dinosaur National Monument Quarry*. Dinosaur Nature Association, Vernal, UT.

Creationist Bibliography—Dinosaurs

1. Armitage, Mark. 2001. "Scanning Electron Microscope Study of Mummified Collagen Fibers in Fossil Tyrannosaurus rex Bone." *CRSQ* 38:61-66.
2. Baker, Mace. 2001. *The Real History of Dinosaurs*. New Century Books, Redding, CA.
3. Baugh, Carl and Wilson, Clifford. 1987. *Dinosaur-2nd Ed*. Promise Publishing, Orange, CA.
4. Bergman, Jerry. 2009. "The Evolution of Dinosaurs: Much Conjecture, Little Evidence." *CRSQ* 46:119-125.

5. Brown, Walt. 2008. *In the Beginning: Compelling Evidence for Creation and the Flood,* Center for Scientific Creation, Phoenix, AZ.
6. Chittick, Donald E. 2007. *The Puzzle of Ancient Man—Evidence for Advanced Technology in Past Civilizations.* Creation Compass, Newberg, OR.
7. Cooper, Bill. 1995. *After the Flood: The Early Post-flood History of Europe Traced Back to Noah,* New Wine Ministries, U.K.
8. Davis, Buddy et al. 1998. *The Great Alaskan Dinosaur Adventure.* Master Books, Green Forest, AR.
9. DeYoung, Donald. 2000. *Dinosaurs and Creation.* Baker Books, Grand Rapids, MI.
10. Doyle, Shaun. 2011. "Cladistics, evolution and the fossils." *Journal of Creation* 25 (2): 32–39.
11. Graham-Kennedy, Elaine. 2006. *Dinosaurs—Where did they come from and where did they go?* Pacific Press Publishing Association, Nampa, ID.
12. Ham, Ken. 1998. *The Great Dinosaur Mystery Solved!* Master Books, Green Forest, AR.
13. Isaacs, Darek. 2010. *Dragons or Dinosaurs.* Bridg-Logos, Alachua, FL.
14. Judkins, Aaron. 2009. *The Global Phenomenon of Human Fossil Footprints in Rock.* Maverick Publishing, Granbury, TX.
15. Lyons, Eric and Butt, Kyle. 2008. *The Dinosaur Delusion—Dismantling Evolution's Most Cherished Icon.* Apologetics Press, Montgomery, AL.
16. Kline, Otis E. and Beh, T.P. 2007. "Rare Dinosaur Excavated by FACT." *Creation Matters,* No. 2 March/April, Creation Research Society.
17. Mitchell, J.D. 2010. *The Creation Dialogues.* WinePress Publishing, Enumclaw, WA.
18. Mitchell J.D. 2013. *Discovering the Animals of Ancient Oregon.* Leafcutter Press, Bellevue, WA.
19. Morris, John D. "Paluxy River: The Tale of the Trails." *Acts & Facts* Vol. 42 No. 5 ICR, p. 12.
20. Oard, Michael. 2011. *Dinosaur Challenges and Mysteries.* Creation Book Publishers, Powder Springs, GA.
21. Oard, Michael. 2003. "In the Footsteps of Giants." *Creation* Vol. 25 No. 2, March-May, p.10.
22. O'Donnell, Philip. 2006. *Dinosaurs Dead or Alive?* Xulon Press.
23. Reed, John K. and Oard, Michael J. 2006. *The Geologic Column—Perspectives within Diluvial Geology.* Creation Research Society, Chino, CA.

24. Ross, Marcus. 2010. "Dinosaurs—Living Large." *Answers* Vol. 5 No. 1, Jan-Mar, p. 26.
25. Swift, Dennis. 2006. *Secrets of the Ica Stones and Nazca Lines.*
26. Taylor, Joe. 1999. *Fossil Facts & Fantasies.* Mt. Blanco Publishing Co., Crosbyton, TX.
27. Thomas, Brian. 2013. "DNA in Dinosaur Bones?" *Acts & Facts* Vol. 42 No. 1, p. 15.
28. Thomas, Brian. 2013. "Original Animal Protein in Fossils." *Creation* Vol. 35 No. 1, p. 14.
29. Ussher, James. 2003. *The Annals of the World.* Master Books, Green Forest, AR.
30. Woodmorappe, John. 1996. *Noah's Ark: A Feasibility Study.* Institute for Creation Research, Dallas, TX.
31. Woodmorappe, John. 1999. *Studies in Flood Geology.* Institute for Creation Research, Dallas, TX.

Glossary: Dinosaur Terminology

AAPS: Association of Applied Paleontological Sciences.

Anterior: Toward the front end of an animal.

Articulated: **Fossil** bones found in the same orientation as they were when the animal was alive are articulated.

Behemoth: A large land animal described in the Bible in Job chapter 40 whose description matches that of a Long-Necked **dinosaur** like *Brachiosaurus*.

Biblical creationist: One who studies God's General Revelation (the creation) while accepting the Bible as being God's accurate and inerrant Special Revelation.

Biblical paleontology: The study of **paleontology** strictly applying biblical **presuppositions**.

Biology: The **science** that studies plants and animals.

Bipedal: Using two limbs for walking, like people, birds, and some **dinosaurs**.

Bone bed: A portion of a **sedimentary rock** layer that has a large number of **fossil** bones or fragments.

Catastrophist: One who believes that many features of the earth are the result of sudden widespread catastrophes rather than gradual evolutionary processes.

Caudal: **Vertebrae** of the tail of an animal.

Cervical: **Vertebrae** of the neck of an animal.

Chevron: A v-shaped bony arch connected to the base of the **caudal vertebrae**, designed for the attachment of tail muscles and to protect nerves and blood vessels.

Cladistics: A method of classification in which life-forms are placed into **taxonomic** groups when they share characteristics (known as **homologies**) that are thought to indicate common ancestry.

Cold-blooded: Same as **ectotherm**.

Coprolite: Trace **fossils** of animal fecal droppings.

Correlation: In **geology** and **paleontology**, the demonstration, or attempted demonstration, of the equivalence of two or more geologic or paleontological phenomena in different geographic areas. **Fossils** are the usual method used to correlate rocks in different areas of the earth.

Creation science: Using biblical **presuppositions** as the basis for interpreting the evidences of the universe.

Cryptozoology: The study of rare animals whose very existence is in dispute.

DNA: Deoxyribonucleic acid, which is constructed by combining nucleotides in long chains. Most genes consist of many thousands of nucleotides. DNA is the genetic material of living organisms and contains the information for life.

Dentary: The bone in the lower jaw that forms the side of the chin.

Dermal: Relating to the skin of an animal.

Dimorphism: The existence of two different types of individuals within a **species**. One example is sexual dimorphism in animals, in which the two sexes differ in size, color, markings etc.

Dinosaur: All of the kinds of animals that were either **ornithischian** or **saurischian**.

Disarticulated: Fossil bones found separated from their orientation when the animal was alive.

Dorsal: To the back of an animal.

Ectotherm: An animal that receives most of its body heat from its environment.

Endotherm: An animal, like a bird or mammal, that obtains most of its body heat from its own metabolism.

Glossary: Dinosaur Terminology

Evidence: Something that makes another thing clear or able to be interpreted.

Evolution: Various models of origins that assume that all organisms on earth descended from a common ancestor by natural processes.

Fact: A reality; truth.

Family: Rank in standard **taxonomy** that lies below **order** and above **genus**.

Femur: The thigh bone of **vertebrate** animals.

Fibula: The smaller and outer of the two bones (other **tibia**) between the knee and the ankle in **vertebrate** animals.

Fossil: Remains or traces of life usually found embedded in **sedimentary rock**. From the Latin word *fossilis* which means "dug up."

Fossil graveyard: Bone bed.

Gastrolith: A stone in the stomach of an animal deliberately swallowed to aid in digestion.

Genus (plural **genera**): Rank in standard taxonomy that lies below the **family** group and above the **species**.

Geologic column: A mental abstraction originally devised to attempt to explain the earth's **geology** but now used to attempt to demonstrate organic evolution in the rock record.

Geology: The study of the earth including rocks, minerals, and its surface and internal processes. Attempts to forensically determine the history of the earth are called "historical geology."

Gizzard: The part of the stomach of some animals in which food is broken up by the action of muscles and possibly **gastroliths**.

Homology: Similarities in the structures of living things, for example the flippers of seals and the hands of people.

Humerus: The long bone of the upper arm.

Hypothesis: An unproved theory, tentatively accepted to explain certain facts.

Ichnology: The study of trace fossils, especially of **fossil** footprints.

Ichnotaxonomy: Taxonomy developed for and applied to trace **fossils**.

Igneous rock: Rock formed by the cooling of molten material. Some rocks crystalize underground (e.g. granite), and others at the surface (e.g. basalt).

Ilium: The largest of the three bones that make up each half of the **pelvic girdle** and are connected to the **sacral vertebrae**. The ilium, the **ischium**, and the **pubis** form the socket for the **femur**.

Index fossil: The **fossil** of an organism believed to have lived during a narrow, well-defined interval of geologic time, and used for **correlation** of rock bodies. In practice, most index **fossils** are marine **invertebrates**.

Interpretation: An explanation using certain **presuppositions**.

Invertebrates: Animals without backbones.

Ischium: Bone of the **pelvis** that is directed backwards.

KT boundary: In secular **geology**, the boundary between the Cretaceous period of the Mesozoic era, and the Tertiary period of the Cenozoic era. By secular definition it also demarcates the time of the extinction of the **dinosaurs**.

Kind: A biblical category of life. God created according to kinds.

Lateral: Toward the side of an animal.

Mammal: Any of a class (Mammalia) of **vertebrate** animals that nourish their young with milk secreted by mammary glands.

Mandible: Lower jaw, comprised of **dentary** bones.

Manus: Hand or front foot.

Mass extinction: The sudden death of a large number of animal groups. The Genesis flood was the greatest of all mass extinction events.

Matrix: The rock or sediment in which a **fossil** is embedded.

Maxilla: One of a pair of bones that form much of the skull anterior to the braincase and hold all of the upper teeth except the incisors.

Medial: Direction toward the middle or inside of an animal.

Megatracksite: A geographical location with large numbers of **fossil** tracks.

Metacarpals: Bones of the hand or front foot.

Metatarsals: Bones of the hind foot.

Morphology: In **biology**, the study of form and structure.

Mutation: A change in the genetic makeup of an organism due to exposure to things such as chemicals or radiation. Almost all mutations are harmful to any organism and evolutionists are yet to demonstrate any reality to their belief that mutations drive organic evolution. See also **natural selection**.

Glossary: Dinosaur Terminology

Naturalism: The philosophy that nature (matter/energy) is the only reality, and that everything in the universe can be explained in those terms without resource to the supernatural. In the secular world "**science**" and "**naturalism**" are synonyms.

Natural selection: The mechanism proposed by Darwin to explain his concept of evolution. It describes the adaptation of organisms to their environment and explains the resulting degrees of survival and reproductive success. Creationists would allow for natural selection (first proposed by a creationist) to affect variability within a kind limited by the information originally placed in the **DNA**. Natural selection would work to reject mutations in order to save the organism rather than use the harmful mutations for progressive evolutionary improvement of some sort. The description of the process as "natural selection" is flawed in the sense that nature (or environment or ecology) is not intelligent, and therefore incapable of "selecting" anything. The creationist would emphasize that adaptation of the organism to its environment is due to innate abilities of the organism, while the evolutionist must attribute intelligence to inanimate things like nature, environment, climate etc.

Opisthotonic pose: The pose a dead animal assumes with its head thrown backwards, its body arched, and its tail arched upwards. Numerous **dinosaur** skeletons have been found in this pose.

Order: Rank in standard **taxonomy** that lies below class and above the **family**.

Ornithischia: A grouping of **dinosaurs** with a pelvic structure similar to that of birds. The word means "bird-hipped."

Paleontology: The study of plant and animal life from the past, including **fossils** found in the rock record.

Paradigm: A way of looking at a particular phenomenon; more encompassing than a theory or model, although sometimes used synonymously with those terms.

Pelvic girdle (Pelvis): The structure in **vertebrates** to which the posterior limbs are attached. It is made up of two halves, each produced by the fusion of the **ilium**, **ischium**, and **pubis**.

Permineralization: The addition of minerals to a bone during fossilization. Not all dinosaur bones are permineralized.

Pes: Hind foot of an animal.

Posterior: Toward the tail end of an animal or bone.

Presupposition: Something supposed or assumed beforehand. The accumulated presuppositions of a person are the foundation for his/her **worldview**.

Pubis: Bone of the **pelvis** that is directed forward.

Quadrupedal: Using four limbs for walking.

Radiometric dating or radioisotope dating: Measuring amounts of atomic isotopes in rocks or other objects, and then assuming that this data gives the sample age.

Radius: One of the two bones that form the lower arm (other is the **ulna**).

Reptile: An **ectothermic**, usually egg-laying **vertebrate**, with an external covering of scales or horny plates. Snakes, lizards, and turtles are reptiles. **Dinosaurs** used to be considered reptiles, but that way of thinking is not currently universally held by paleontologists.

Rostral: The bone that forms the upper beak in Horn-Faced **dinosaurs**.

Sacral vertebrae: Vertebrae that are fused together and support the **pelvic girdle**.

Saurischia: A grouping of **dinosaurs** with a **pelvic** structure similar to that of modern **reptiles**. The word means "lizard-hipped."

Sauropods: The **quadrupedal** Long-Necked **dinosaurs**.

Scapula: Portion of the shoulder blade that forms part of the socket for the **humerus**, along with the coracoid.

Science: The systemized knowledge derived from observation and study. True science is not **naturalism**.

Secularism: The anti-religious view that true knowledge can only be found through rational or empirical means. Secularists live their lives with an orientation toward the present world only, as opposed to living from the biblical perspective that life is everlasting.

Sedimentary rock: Rock layers formed when sediment settles in water, and then hardens later. **Fossils** are usually found in sedimentary rock.

Species: A taxonomic group into which a **genus** is divided.

Stomach stone: A **gastrolith**.

Stratigraphy: The study of rock strata, concerned with the characters and attributes of rocks and their interpretation in terms of mode of origin and geological history.

Glossary: Dinosaur Terminology

Taphonomy: The study of the conditions and processes by which organisms are fossilized and preserved.

Taxonomy: The orderly arrangement of animals and plants according to their presumed natural relationships. Classical taxonomy's hierarchical classifications are (in descending order) kingdom, phylum, class, **order, family, genus,** and **species**. For evolutionists, the relationships are assumed to indicate that all life has a common ancestor.

Tectonics: Large scale deformation of the earth's crust.

Theropods: The **bipedal saurischian dinosaurs**.

Tibia: Primary bone of the lower hind leg. See also **fibula**.

Trace fossil: Fossilized tracks, trails, burrows, droppings, or tubes resulting from the life activities of animals.

Ulna: One of the two bones of the lower arm. The upper end of the ulna forms the elbow. See also **radius**.

Uniformitarianism: The view that geological and biological changes in the past always occurred at the slow rate often measured today. The principle is regularly stated as "the present is the key to the past."

Ventral: Pertaining to a surface at the interior of an animal.

Vertebra (plural vertebrae): A bone of the backbone. The vertebrae together support the body and protect the spinal cord.

Vertebrates: Animals with backbones.

Warm-blooded: Same as "**endotherm**."

Worldview: The way a person looks at the world, or his perception of reality based on his basic **presuppositions** about what is true. Every worldview, whether expressly religious or supposedly non-religious, is a belief system that begins with **presuppositions** (assumptions) held by faith.

Index

A

Abramson, Paul 32
Academy of Natural Sciences of Drexel University, The 57, 74
Achelousaurus ix, 112
Acrocanthosaurus ix, 55, 59, 194, 217
Adam and Eve xxi, 10, 17
Alamosaurus ix, 73, 161
Albertosaurus ix, 51, 58, 194, 207, 210-211
Alley Oop 8-11
Allosaurus ix, 19, 22, 53-54, 56-57, 60-63, 73, 194, 196-200, 206, 214-215, 220
American Museum of Natural History 22, 33, 56, 71-73, 75, 142
Ammosaurus ix, 74, 145
Anatotitan ix, 56, 98, 105
Anchiceratops ix, 72, 112
ancient man 10
Anchisaurus ix, 74-75, 145
Andrews, Roy Chapman 71-72
Animantarx ix, 61, 82-83
Ankylosaurus ix, 95, 97
Answers in Genesis 55, 65, 73
Anterior 18
Apatosaurus ix, 54, 56-57, 75, 161, 163-165, 167
Appalachiosaurus ix, 194
Archaeopteryx 63

Armor-Backed **82-87**
armor plate 82, 95, 112
Arrhinoceratops ix, 112
Association of Applied Paleontological Sciences 67
asteroid theory 46
Avaceratops x, 57, 112

B

Babel 10
Badlands 69
Bakker, Robert T. 145
Barosaurus x, 52, 56, 75, 161, 166, 174
Baugh, Carl 36
Behemoth 5, 10
Bender, Jack and Carole 11
Bible xx-xxi, 1-6, 8, 10, 14, 33, 46, 65, 79-80
biblical paleontology xx, 7, 26
biblical timescale 1
bipedal 16-17, 80
Bipedal Browser **88-94**
Birds 3, 5-6, 10-11, 13, 17, 19, 38, 176
Bird, Roland T. 33
Black Hills Institute Museum 58
bone beds 22, 46-47
bone cabin 23-25
bone wars 66, 72, 75
Boylan, Thomas 23
Brachiosaurus x, xv, 54, 149, 153-154

235

Brachyceratops x, 73, 112
Brachylophosaurus x, 98, 102, 104
brain cast 191, 213
Brigham Young University
  Museum of Paleontology 60
Brontosaurus 11, 22-23, 28, 75, 161
Brown, Barnum 71-72, 76, 95, 137

C

Camarasaurus x, 20, 22, 54, 73, 75, 149-150, 152, 155-157
Camp, Charles Lewis 142
Camposaurus x, 142
Camptosaurus x, 53-54, 60-61, 75, 88-90
Canadian Museum of Nature 52
Carnegie Museum of
  Natural History 57, 73, 134
Cedar Mountain 149
Cedarosaurus x, 149
Centrosaurus x, 74, 112, 117
Ceratosaurus x, 134-136
Chasmosaurus x, 51, 57, 61, 74, 112, 114-115, 122
Chevron 172
Chide Point 142
Chindesaurus x, 142
Christ 14
Christian Church 5
Cladistics xvii, 81
Cleveland-Lloyd Dinosaur
  Quarry 60-61
Cleveland Museum of Natural
  History 57
Cloverly Formation 140
Club-Tailed **95-97**
Coelophysis x, 53, 56-57, 72-73, 142-144
Coelurus x, 142
Colbert, Edwin H. 72, 142
College of Eastern Utah
  Prehistoric Museum 61
Common Designer xxi
Como Bluff 22-26, 72

Cope, Edward Drinker 22, 66, 72, 74-75, 88, 98, 112, 142, 149
Correlation 8
Corythosaurus x, 52, 56-57, 98
created kinds ix, xvi, xx, 14, 77, 79-220
Creation Evidence Museum 36
Creation Museum 40-41, 51, 55
Creation Science 3, 6, 241
Cryptozoology 13
Curse xxi, 14

D

Daspletosaurus x, 52, 54, 194, 203
Deinonychus x, 54, 56-57, 59, 137, 140
Delk, Alvis 35-36
Delk Track 35-36
Denver Museum of Nature and
  Science 53
Diceratops x, 112
Dilophosaurus x, 53, 142
dinosaur anatomy xix, 18
dinosaur death 45-50
dinosaur digs 65-68
dinosaur eggs 13, 37-44
dinosaur extinction 4, 45-46
dinosaur hipbones 14-17, 80
Dinosaur National Monument 26, 53, 73
Dinosaur Provincial Park 76
dinosaur skin xix, 11, 106-107, 117, 131-132
dinosaur skulls (See "Skulls")
dinosaur tracks (See "footprints, dinosaur")
Dinosaur Valley State Park 30, 33
Diplodocus x, 22, 53-54, 57, 59, 62, 161-162, 164-168, 175
DNA 6, 8, 14, 50, 79
Douglass, Earl 73
Dragons 4-5, 10
Drinker x, 88
Dromaeosaurus xi, 55, 137, 141
Dromiceiomimus xi, 176
Duck-Billed xviii, **98-111**

Index

E

Ectothermic	13
Edmarka	xi, 194
Edmontonia	xi, 82
Edmontosaurus	xi, xv, 20, 51, 58-59, 98, 103, 106-110
egg fossils	37-44
Einiosaurus	xi, 112
Elephant	5, 11
Embryos	38-39
Endothermic	13
Eobrontosaurus	xi, 161
Eucoelophysis	xi, 142
Euoplocephalus	xi, 51, 95-96
Evolution	xv-xvii, xx, 5-6, 8, 11, 14, 36, 50, 72, 75
extinction theories	4, 45-46

F

Fall	xxi, 3, 10, 14, 80
Femur	125, 152-153, 155, 157, 163, 166, 184-185, 198, 200-201, 214
Fibula	110, 155, 171
Field Museum of Natural History	54
footprints, dinosaur	27-33, 74
footprints, human	32-36
Fort Worth Museum of Science and History	58-59
Fossil Cabin	23
fossil collecting	65-69
fossilization	48-49
Foundation Advancing Creation Truth	65
Furcula	19, 218-220

G

Garden of Eden	14
Gastonia	xi, 60-61, 63, 82, 84, 87
Gastroliths	150-151, 161, 176, 180
Genealogy	1-2
Genesis	xxi, 1-3, 5-6. 10, 26, 46, 79
Genesis Flood, The	33
Genetics	xxi, 8, 14, 50

genus names	ix, 19, 79
Geological Survey of Canada	74, 76
Ghost Ranch	72, 142
Giles, Brent	32
Gilmore, Charles W.	73, 189
Glendive Dinosaur and Fossil Museum	51, 55, 66
Glen Rose, Texas	29-30, 33, 36
God	xx-xxi, 1-6, 10, 13-14, 16-17, 19, 26, 38, 67, 79
Gojirasaurus	xi, 142
Gorgosaurus	xi, 51, 59, 194, 204, 207, 218
Grallator	28
Gryposaurus	xi, 98, 104

H

Hadrosaurus	xi, 98
Hamlin, V.T.	9
Hell Creek Formation	123-124, 192, 208-213
Hesperosaurus	xi, 180, 186-188
Hippopotamus	5
Homology	xx
Horn-Faced	xviii, 19, **112-133**
Horn-Nosed Bipedal	**134-136**
Houston Museum of Natural Science	59
Humerus	154
Hypacrosaurus	xi, 98

I

Ichnology	27-28, 31
Ichnotaxonomy	28
Ilium	14
Indiana Jones	71
Ischium	14

J

Job	5
Judith River Formation	141, 211
Jurassic Park	137

K

K/T boundary	46
Kayenta Formation	143
Killer-Clawed	**137-141**
King David	10
Kritosaurus	xi, 99

L

Lambe, Lawrence	74, 82, 95, 109
Lambeosaurus	xi, 51, 74, 99-101, 103-105
Lance Formation	179, 193
Lateral	18
laws and regulations	67
Leidy, Joseph	74, 98, 137
Leptoceratops	xi, 112
Leviathan	10
Lithe, Fast Running	73, **142-144**
Long-Necked Big-Clawed	**145-148**
Long-Necked Boxy-Headed	**149-160**
Long-Necked Slender-Headed	11, **161-175**
Lull, Richard S.	74

M

Maiasaura	xi, 63, 99, 108, 111
Mammals	13, 22, 38, 45
Marsh, Othniel Charles	22, 66, 74-75, 88, 98, 112, 134, 145, 161, 176, 180, 194
Marshosaurus	xi, 194
Megatracksites	29
Metatarsal	167
Meyer, Grant	137
Mokele-mbembe	13
Monoclonius	xi, 112, 118-119
Montanoceratops	xi, 113
Morris, Henry	33
Morrison Formation	85-86, 156-157, 165
Mount St. Helens	8
Museum of Northern Arizona	53, 72
Museum of the Rockies	39, 42-43, 56
Museums	xviii, 51-63

N

Nanotyrannus	xi, 57, 194, 212
National Geographic	33
National Museum of Natural History Smithsonian Institution	22, 62-63, 73, 134
Natural History Museum of Los Angeles County	53
Naturalism	6, 8
Nests	38, 42-43
Niobrarasaurus	xi, 82
Noah	xxi, 10
Noah's ark	xx, 3-4, 10, 79

O

Oard, Michael	31
Ojo Alamo	161
opisthotonic posture	48-49
ornithischian	14-16, 80
Ornitholestes	xii, 51, 142
Ornithomimus	xii, 51, 176-178
Orodromeus	xii, 88
Osborn, Henry Fairfield	74-75, 176, 194
Ostrich-Like	**176-179**
Ostrom, John	137
Othnielia	xii, 53, 63, 88, 90-91
Owen, Sir Richard	5

P

Pachycephalosaurus	xii, 55, 57, 189, 191-193
Pachyrhinosaurus	xii, 113, 116
Paleontology	xv, xvii, xx, 6-7
Paluxy River	29-30, 32-34
Paluxysaurus	xii, 58, 149, 158-160
Panoplosaurus	xii, 82
Parasaurolophus	xii, 52, 54, 99
Park, William	74, 88
Parkosaurus	xii, 88
Pawpawsaurus	xii, 82
Pentaceratops	xii, 57, 113
Permineralization	48
Plate-Backed	**180-188**

Index

Plateosaurus xii, 75, 145-148
Podokesaurus xii, 142
Posterior 18
Presuppositions xix-xx, 1, 5-6, 31, 33, 36
Prosaurolophus xii, 51, 61, 99-101
Pubis 14, 160, 206
Purgatoire River Trackway 29

Q

Quadrupedal 17, 80

R

radiocarbon dating 8
radiometric dating 7-8
Red Deer River 72, 74
Reptiles 13, 38, 54
Ribs 172
Riggs, Elmer G. 149
Royal Ontario Museum 52
Royal Tyrrell Museum of Paleontology 51-52

S

Sam Noble Oklahoma Museum of Natural History 57
Saurischian 14-16, 80
Saurolophus xii, 99
Sauronitholestes xii, 51, 137, 139
Sauropelta xii, 82
Saurophaganax xii, 57, 194
Sauropod 17-18, 30, 149
Sauroposeidon xii, 149
Scapula 126, 153
Segi Canyon 142
Segisaurus xii, 142
Seismosaurus xii, 161
Silvisaurus xii, 82
Skulls 19-22, 80, 86, 92, 101-106, 109, 114, 118-126, 130, 135, 140-141, 144, 151, 156-158, 164-165, 179, 182, 187, 190-193, 198-199, 201. 205-210, 212

soft tissue xvii, 8, 50
Sphaerotholus xiii, 189
Stegoceras xiii, 189
Stegosaurus xiii, 21-22, 52-57, 59, 62, 73, 75, 180-186
Sternberg, Charles H. 76, 82
Stokesosaurus xiii, 194
Stone Age 10
Struthiomimus xiii, 58, 176, 179
Stygimoloch xiii, 189
Styracosaurus xiii, 74, 113
Sue (*T-rex*) 54, 194, 212
Supersaurus xiii, 63, 161, 168-173

T

tail spike 180, 184, 187
Taphonomy 47
Taxonomy xvi-xvii, xx, 17
Taylor, Stan 34
Taylor Trail, The 34-35
Teeth xvi, 11, 18-19, 47, 80, 118, 137, 150, 161, 180, 195, 200, 214
Tenontosaurus xiii, 57-58, 88, 92-94
Theropod 17-18
Thescelosaurus xiii, 58, 73, 88
Thick-Headed **189-193**
Tibia 102, 110, 155, 171
tools, fossil hunting 70
Torosaurus xiii, 113, 121
Torvosaurus xiii, 60, 194, 197
trace fossils 27
tracks 27-31, 35-36
trackways 27-29, 34
Triceratops xiii, xv, 21, 51-53, 55-56, 58-59, 62-63, 71, 75, 112, 114-116, 119-120, 122-127, 130-133
Troödon xiii, 137
Two Medicine Formation 207, 210
Turtles 22
Tyrannosaurus
  xiii, xv, 51-59, 62-63, 71, 194, 198, 200-206, 208-213, 215-216, 219
Tyrant Bipedal xviii, 11, 19, **194-220**
Tyrrell, J.B. 194

U

Ulna 125
Uniformitarianism 8, 47, 67
University of Alberta 52
Utah Fieldhouse of Natural History
　State Park Museum 62
Utahraptor xiii, 61, 137-139

V

Variation xvi-xvii, xxi, 5, 14, 18-19, 79-80
Velociraptor 137
Ventral 18
Vertebrae 153, 157, 165, 168-170, 173-174
volcanic eruptions 4, 46

W

Woodmorappe, John 3
worldwide flood xx-xxi, 2, 4, 10, 14, 26, 31, 46-50
Wyoming Dinosaur Center 63

Y

Yale Peabody Museum
　of Natural History 22, 54, 74-75

Z

Zallinger, Rudolph F. 54
Zuniceratops xiii, 63, 113, 127-128

J.D. Mitchell is a registered professional engineer in Oregon and Washington and has a Bachelor of Science in Mechanical Engineering from the University of Washington. He has also completed his Master of Biblical Studies in Biblical Creation Apologetics from Master's Graduate School of Divinity.

He is principal consulting engineer for Crane & Hoist Engineering of Gresham, OR, and is Executive Director of the Institute for Creation Science (www.icspdx.org) that meets monthly in Portland, OR. He does creation science research, writes and speaks regarding the creation versus evolution controversy as a part of his creation ministry Creation Engineering Concepts (www.creationengineeringconcept.org). He is also a member of the Design Science Association of Oregon, the Creation Research Society and the American Society of Mechanical Engineers. He is listed in the *Marquis Who's Who in Science and Engineering.*

J.D. has been studying the scientific and biblical evidence regarding the origins controversy since 1984 when he was converted from a theistic evolutionist to a born again Christian. He has had a prodigious interest in dinosaurs since he was nine years of age.

Guidebook to North American Dinosaurs According to Created Kinds is his third book, and follows *The Creation Dialogues* (WinePress Publishing) that came out in 2010 and *Discovering the Animals of Ancient Oregon* (Leafcutter Press) that was published in 2013.

CPSIA information can be obtained at www.ICGtesting.com
Printed in the USA
LVOW01s2123060715

445150LV00012B/315/P